Into The Jungle

GREAT ADVENTURES IN THE SEARCH FOR EVOLUTION

Sean B. Carroll

WITH ILLUSTRATIONS BY

LEANNE M. OLDS AND JAMIE W. CARROLL

PEARSON

Benjamin
Cummings

SAN FRANCISCO   BOSTON   NEW YORK   CAPE TOWN   HONG KONG   LONDON   MADRID
MEXICO CITY   MONTREAL   MUNICH   PARIS   SINGAPORE   SYDNEY   TOKYO   TORONTO

Editor-in-Chief: Beth Wilbur
Executive Director of Development: Deborah Gale
Senior Development Editor: Susan Teahan
Editorial Assistant: Logan Triglia
Executive Managing Editor: Erin Gregg
Managing Editor: Michael Early
Production Supervisor: Camille Herrera
Production Service and Composition: Progressive Publishing Alternatives
Interior Designer: Seventeenth Street Studios
Cover Designer and Illustrator: Leanne Olds
Cover Production: Mark Ong
Senior Art Editor: Donna Kalal
Manufacturing Buyer: Michael Penne
Director of Marketing: Christy Lawrence
Executive Marketing Manager: Lauren Harp
Cover printer: Phoenix Color

Some of the content of this book is also being published by Houghton Mifflin Harcourt for the trade market in *Remarkable Creatures: Epic Adventures in the Search for the Origins of Species* by Sean B. Carroll

Library of Congress Cataloging-in-Publication Data

Carroll, Sean B.
    Into the jungle : great adventures in the search for evolution / Sean B. Carroll.
       p. cm.
    Includes bibliographical references and index.
    ISBN 0-321-55671-2
 1. Evolution (Biology)—History.   2. Naturalists—Travel—History.   I. Title.
    QH361.C27 2009
    576.8—dc22   2008021635

ISBN 10: 0-32155-671-2   Component ISBN 10: 0-32159-848-2
      13: 9-78032155-671-4   13: 9-78032159-848-6

         13 14   15 14

www.pearsonhighered.com

About the Author

Sean B. Carroll is Professor of Molecular Biology and Genetics and an Investigator at the Howard Hughes Medical Institute at the University of Wisconsin–Madison. Among the most prominent biologists working in the world today, Carroll is a member of the National Academy of Sciences, an award-winning author of two highly acclaimed books on science for the general public (*Endless Forms Most Beautiful* and *The Making of the Fittest*), a widely known charismatic public speaker, an ardent advocate for science education, and a frequent guest on *NOVA* and other popular television and radio programs. He has authored or co-authored two textbooks and more than 100 scientific papers on animal development and evolution.

For my family, the greatest adventure of all.

Contents

Into The Jungle

Preface

If history were taught in the form of stories, it would never be forgotten.

—Rudyard Kipling, author of *The Jungle Book*

Not so long ago, most of the world was an unexplored wilderness. The animals, plants, and people (if any) that inhabited the land beyond Europe were unknown, at least as far as the Western world was concerned. The rivers and jungles of the Amazon, the Badlands of Patagonia and of the American West, the tropical forests of Indonesia, the African savannah, Inner Mongolia, the Southern Ocean, and the Antarctic were complete mysteries.

The explorations of these marvelous places are some of the epic adventures of recent times, and the discoveries they yielded gave birth to and nourished the science of evolutionary biology. In this short book, I will tell the stories of some of the greatest adventures in and discoveries of biology.

We will visit six continents and encounter many amazing creatures of the past and present. We will also meet some remarkable people. The characters in these stories followed their dreams — to see exotic places, to collect and study previously unknown species, or to unearth the past. Few, if any, had some notion of great achievement or fame. They were driven by a passion for natural history, and they were willing, sometimes eager, to take great risks. Many faced, and experienced, the perils of traveling long distances by sea. Many left behind skeptical and anxious loved ones. Once arriving safely at their destinations, they encountered all sorts of dangers — earthquakes, bandits, hostile natives, wars, poisonous snakes, nasty insects, and many deadly diseases.

Their triumphs were much more than survival, and the collecting of specimens of what the world holds. The pioneers in these stories, provoked by a riot of diversity beyond their wildest imaginations, turned from collectors into *scientists*. They posed, and began to answer, the most fundamental questions about Nature. They asked not just what existed, but how and why it came to be. Scientists have ever since been walking literally and figuratively, in their footsteps.

As you read these stories, just imagine. Imagine what it would be like, at or about your age, to travel to a far-off unexplored land. Not for just a couple of weeks, or a month, or even a year, but for five, eight, or even ten years. No family, no comforts of home. All that lay ahead was unknown, but also unexplored, for you to discover.

Imagine the joy in finding an iridescent butterfly larger than most birds, or a nest of dinosaur eggs, or a fish thought to be long extinct, or the skull or leg bone of an ancient human. The naturalists in these stories found these wonders, and much more. Most importantly, these discoveries opened their eyes, and in turn open our own, to new ways of seeing and understanding Nature.

* * *

I wrote this book for several reasons and with a wide variety of students in mind. My primary goal is to provide an altogether different way of learning about biology, and evolutionary biology in particular. Textbooks are packed with up-to-date information, but great achievements may be compressed into just a few lines that do not convey the drama and richness of the process of scientific discovery. Those discoveries are far more interesting and better understood when we walk in the shoes of those who are searching for clues and see how they pieced them together into a bigger picture.

I am convinced that Rudyard Kipling, who knew a thing or two about writing memorable stories, was right — that history is far more memorable when told in the form of stories. I am certain that science is no different. To tell, and to want to hear stories, is a big part of being human. From the rock and cave art of our ancestors to the ancient Greeks, Shakespeare, Dickens, modern music, and movies, we have shared our journeys, experiences, and pictures of the world through various forms of storytelling.

You may wonder, "Why are all of these stories about evolution?" I'll borrow one explanation from paleontologist George Gaylord Simpson, one of the major figures in evolutionary science in the 20th century, who stated, "Life is the most important thing in the world, the most important thing about life is evolution." Or as it has also been described: evolution is the greatest show on Earth. An understanding of that show, and all that it means when grasping our place in the world, is the most important gift biology has to offer.

I wrote these stories with nonscience students particularly in mind. I realize full well that most biology students do not become biologists or doctors, and that the vast body of knowledge covered in a typical course begins to fade as soon as the course is over (or sooner!). I hope that these stories will resonate differently than textbook material, will stick with you a bit longer, and that they will appeal as much to future English teachers, artists, and politicians as they may to the next generation of paleontologists. And maybe, if some strike a chord, they might pique enough interest in Darwin's life, in the Amazon, Africa, the Antarctic, human pre-history, or any aspect of natural history, that you might want to look into some of the suggestions for further reading I have given in the back of the book.

But above all, I hope that you enjoy them.

Sean B. Carroll
Madison, Wisconsin

For more on the story in any chapter, go to www.aw-bc.com/carroll
for journal articles, analysis and discussion questions, additional photos,
and hyperlinks to additional sources.

PART I

Into the Jungle

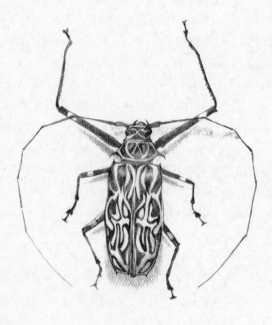

Three great voyages and three great naturalists were center stage in the conception of the idea of natural selection, and the early development of evidence in support of it. Best known of all is, of course, Charles Darwin and the voyage of the HMS *Beagle* and whose name is forever linked to places such as the Galapagos Islands. While Darwin is justifiably famous for his contributions to natural history and the theory of evolution, two other naturalists, Alfred Russel Wallace and Henry Walter Bates, tend to be underappreciated, or neglected altogether. But Wallace and Bates undertook great adventures, much longer in duration and under even more difficult circumstances than Darwin's, and they developed ideas and evidence relating to the mechanisms of evolution. Although Darwin's journey (1831–1836) preceded those of Wallace (1848–1862) and Bates (1848–1859) by about twenty years, their work became intertwined upon Wallace's and Bates' return to England. The three adventurers were friends and correspondents for the rest of their lives.

All three were young men when they set out from England for the jungles of South America. Darwin was twenty-two, Bates was twenty-three, and Wallace was twenty-five. But unlike Darwin, who was educated at Cambridge, came from a wealthy family, and traveled as ship's naturalist on an armed British vessel, Bates and Wallace were self-taught entrepreneurs. Wallace left school at thirteen and learned several trades and skills. He eventually became a teacher in Leicester, where he met Bates. Bates, too, left school at thirteen and became apprenticed to a hosier. The two were keen insect collectors. Through this passion they developed a friendship and a mutual fascination with the Tropics and the prospects of collecting some of the rich diversity of species found there. Both read avidly, including Darwin's *Voyage of the Beagle.* A then new book by American William H. Edwards, *Voyage up the River Amazon,* and a chance encounter with Edwards in London clinched their decision to go to the Amazon and to pay for their expenses by shipping and selling prized specimens back in England.

After a year and a half together in the Amazon, Wallace and Bates split up to cover more territory. Wallace headed home after four years and then undertook a long, successful expedition to the jungles of the Malay Archipelago. Bates stayed in the Amazon jungle for eleven difficult but rewarding years.

Contrary to what some may believe, or one might expect, Darwin did not set out with any intent of gathering evidence for or against any great idea. His theory of evolution took shape after the voyage as he pondered what he had seen and began to question (privately) the thinking at the time. Wallace and Bates, on the other hand, did have evolution in mind from the start. The idea that species change was percolating in scientific and popular circles and inspired Wallace and Bates to go gather data "toward solving the problem of the origin of species."

The stories here are brief glimpses of the highlights of their adventures and discoveries. The three explorers left behind a rich legacy of memoirs, narratives, correspondence, and scientific work which are readily accessible.

Charles Darwin & his sister Catherine
From a chalk drawing in the possession of
Miss Wedgwood of Leith Hill Place.

FIGURE 1.1 *"I was in many ways a naughty boy."*
Portrait of young Charles and his sister Catherine. Charles later wrote in his
autobiography, "I was much slower in learning than my younger sister Catherine, and
I believe that I was in many ways a naughty boy." *From* More Letters of Charles
Darwin: A Record of His Work in a Series of Hitherto Unpublished Letters edited by
F. Darwin and A. Seward *(D. Appleton and Co., New York, 1903).*

Reverend Darwin's Detour

Every traveller must remember the glowing sense of happiness, from the simple consciousness of breathing in a foreign clime, where the civilized man has seldom or never trod.

—Charles Darwin, *Voyage of the Beagle* (1839)

His nickname was "Gas".

Thirteen-year-old Charles Darwin was, as most younger brothers are, highly susceptible to conspiracy and mischief with his older brother. Erasmus ("Ras"), five years his senior, had developed an interest in chemistry and recruited his little brother into outfitting a makeshift lab in the garden shed of the family home. The two boys pored over chemistry manuals and often stayed up late at night concocting some noxious or explosive mixture.

Sons of a wealthy doctor, Ras and Gas always had plenty of funds available for their hobbies. They bought test tubes, crucibles, dishes, and all sorts of other apparatus. Of course, chemistry wouldn't be as much fun without fire, so the boys invested in an Argand lamp, a type of oil lamp that they used to heat chemicals and gases. Their fledgling laboratory also had fireproof china dishes, courtesy of their Uncle Josiah Wedgwood II, the leading maker of pottery in England.

Charles enjoyed the prestige his stinky shed earned him among schoolmates. He was also very well-liked for his cheerful and mild-mannered disposition. Some joined him on his expeditions into the countryside collecting insects or bird hunting. The boarding school he attended was only a mile from home, so he knew the surrounding woods and streams very well.

The schoolmaster, however, was not so impressed with either Charles' chemistry or his lackadaisical approach to the classics. Charles was not much of a student. He was bored stiff by the rote learning of ancient geography, history, and poetry demanded by the school. He escaped to the woods and home to visit and play with his dog as often as he could, sometimes risking expulsion should he get locked out before bedtime. Racing back to his dormitory, he prayed out loud for God's help in getting him there in the nick of time, and marveled that his prayers were answered.

Charles' father was increasingly aware of his son's dislike for school. Although Charles adored him, Robert Darwin was a large, imposing figure, and what he said went for the Darwin household. "The Doctor," as Robert Darwin was called, was worried that Charles was frittering away his opportunities. One day, The Doctor's anger erupted, "You care for nothing but shooting, dogs, and rat-catching, and you will be a disgrace to yourself and your family!"

Eventually, The Doctor decided that the best thing would be to take Charles out of school two years early, at the age of sixteen, and send him to Edinburgh where he could stay with his brother and enroll in medical school. The Doctor hoped that Charles would follow in his footsteps, and those of his grandfather, and become a physician.

Doctor Darwin?

In Edinburgh, Charles did learn many things — taxidermy, natural history, zoology — and that he did not want to be a doctor.

Edinburgh University provided the best medical training in Great Britain, but it was a gruesome ordeal in the 1820s. Charles was repulsed by the professor of anatomy, who showed up for his lectures dirty and bloody, fresh from stints at the dissecting table. Charles also found surgery sickening. After witnessing an operation on a child (before the advent of anesthesia), Charles fled the operating theatre and vowed not to return.

Outside of the horrors of medical school, Edinburgh did offer some attractive excursions. Charles loved to walk the dramatic coastline of the Firth of Forth and to look for sea creatures that washed ashore. In the city, he met a freed slave from Guiana (on the northeast coast of South America), John Edmonstone, who agreed to tutor Charles on how to stuff and mount birds. Charles was an excellent student and reveled in Edmonstone's tales of the Tropics. Edmonstone's vivid descriptions of the South American rainforests were perfect antidotes for the bone-chilling Scottish climate.

After two years of cutting classes, and drifting about Edinburgh, Charles was no closer to being a doctor, or anything else for that matter. He dropped out of medical school without a degree.

The Doctor would have to find something respectable for his aimless son. There was a great risk among the well-to-do that their privileged sons would be content to live off the family fortune. If not a doctor or lawyer, what would befit Charles? What position would bring him the most respectability for the least ambition?

That would be the Church of England.

It was common practice in Charles' day for parishes to be auctioned off to the highest bidder, who would then install a family member as parson. It was a comfortable lifestyle with ample lodging and some land, along with income from the local parishioners and investments. Charles would have plenty of time to pursue his hobbies.

The only requirement would be to pass his ordination, which required a bachelor's degree from Cambridge or Oxford, and a year's study of divinity. So it was off to Cambridge for Charles, where virtually all of the faculty were ordained members of the clergy.

The Making of a Country Parson

Having washed out of Edinburgh, Charles was determined to start off on the right foot at Cambridge. Unfortunately, his resolve would soon be challenged by the beetle collecting craze that was then sweeping the nation and Cambridge. The capture of diverse and rare species was becoming a competitive sport. It so appealed to Charles' love for romping about the woods with like-minded comrades and his thirst for recognition that he soon became obsessed.

Charles acquired the best equipment, hired helpers to sift through forest debris, and spent significant sums buying specimens from other collectors. One day, peeling bark off a tree, he eyed two rare beetle forms and quickly grabbed one in each hand, but then spotted another. So he popped one of the two captured beetles into his mouth so that he could snag the third. Unfortunately, the one in his mouth was a bombardier beetle, and it emitted an awful brew that forced Charles to spit it out and to lose the other two!

Charles spent most of his first two years pursuing such beetle trophies, but finally resolved again to buckle down and to prepare for the key exam at year's end. He was to be tested on translation of Latin and Greek, portions of the Gospels, the New Testament, and on the works of Reverend William Paley, who had written several books concerning the evidence for God and the truths of Christianity. Charles, in fact, was lodging in the very rooms Paley had occupied at Cambridge decades earlier and was very impressed and persuaded by Paley's lucid logic.

Charles made it through the exams and resumed his beetling, but he also fell under the very positive influence of his professor of botany, Reverend John Stevens Henslow. On Friday nights, Henslow hosted small gatherings at his home for the discussion of natural history, and a little wine drinking. Other professors would occasionally drop in and share their expertise and passions. Charles had found his place. Henslow took Charles under his wing, and the two were so often seen walking together engrossed in conversation that Gas became instead "the man who walks with Henslow."

Henslow took students on all sorts of botanical excursions around Cambridge. Charles was eager to please, even wading through the muck of the

River Cam to snatch a rare species for his mentor. Charles saw Henslow as the role model of the ordained naturalist. Charles said of Henslow, "[He is] quite the most perfect man I ever met with." Charles thought he would take his year of divinity studies under Henslow.

First, there was the matter of passing his final exam — more Homer, Virgil, and Paley — with some math and physics tossed in. Charles ranked tenth in a class of 178. To get his degree, he also had to pledge his adherence to the "Thirty-Nine Articles" (established in 1563), which outlined the basic doctrine of the Church of England.

Henslow continued grooming Charles. He encouraged Charles to read more and to think about traveling to widen his horizons. He lent Charles his copy of Alexander Von Humboldt's *Personal Narrative*. While the old Charles would have struggled with the seven-volume work, the new Charles gobbled it up and started dreaming of the places Humboldt described in his travels to and throughout South and Central America. The Canary Islands (off the coast of Northwest Africa) offered the closest glimpse of a tropical paradise, so Charles thought, "Why not go there?" Henslow and three of Charles' friends were interested in going along. Charles went to see his father, and The Doctor forked over the money to clear all of Charles' debts and to pay for the expedition.

Henslow knew that Charles would need some geological training to make the most of such a visit, so Henslow set up a "hands-on" tutorial for Charles with Reverend Professor Adam Sedgwick, a leading figure of British geology who would later name the Devonian and Cambrian geological periods. Sedgwick, who had also trained Henslow, took Charles on a field trip to Wales. Sedgwick taught Charles very well. Charles discovered that he had a knack for and love of geology.

While he was in Wales, Charles' mates for the Canary Island trip backed out one by one until just a single companion remained. On his way back to England, Charles received a message that this friend had died suddenly. He was shocked over the loss, and deeply disappointed that his planned expedition had dissolved.

When he reached home, exhausted and uncertain of what he was going to do, there was mail waiting for him. It was a letter from Henslow with stunning news. Charles was being offered a voyage around the world.

Permission to Board

Asked to recommend a naturalist for an upcoming voyage to South America, Henslow had written to his young protégé Charles, "I think you are the very man they are in search of."

Charles was elated. The Doctor was not.

Charles' father knew that British ships were sailed by some very rough characters. He also knew that the ships were dangerous and often wound up being sailors' coffins. The Doctor thought the voyage too risky an adventure and yet another delay on Charles' path to settling down in a respectable position. Charles glumly wrote to Henslow that he would not go over the objections of his father.

To distract himself from his disappointment, Charles headed off to his Uncle Josiah's house. His father, however, handed him a letter for his uncle. The Doctor explained that he objected to the voyage on many grounds but told his brother-in-law, "If you think differently from me I shall wish him [Charles] to follow your advice."

Uncle Josiah was indeed much more supportive of the adventure. So that he could offer an informed response to The Doctor, Uncle Josiah asked Charles to write down a list of his father's objections. (The list still exists today, Figure 1.2.) His father's words still fresh in his mind, Charles recited the objections:

1. Disreputable to my character as a Clergyman hereafter
2. A wild scheme
3. That they must have offered to many others before me, the place of Naturalist
4. And from its not being accepted there must be some serious objection to the vessel or expedition
5. That I should never settle down to a steady life hereafter
6. That my accommodations would be most uncomfortable
7. That you should consider it as again changing my profession
8. That it would be a useless undertaking

Uncle Josiah did see it all very differently, and he wrote his response to The Doctor, who quickly changed his mind about the South American voyage. The Doctor declared he would give Charles' journey all the assistance in his power.

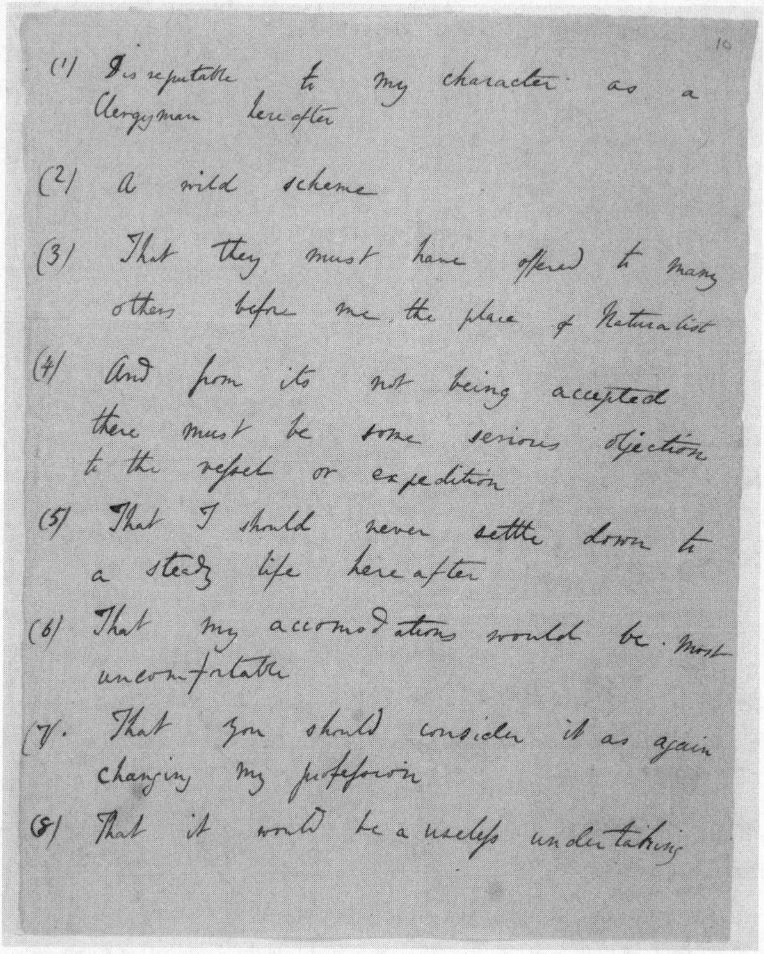

10

(1) Dis reputable to my character as a Clergyman hereafter

(2) A wild scheme

(3) That they must have offered to many others before me, the place of Naturalist

(4) And from its not being accepted there must be some serious objection to the vessel or expedition

(5) That I should never settle down to a steady life hereafter

(6) That my accomodations would be most uncomfortable

(7) That you should consider it as again changing my profession

(8) That it would be a useless undertaking

FIGURE 1.2 *Charles' list of his father's objections to the Beagle voyage. Reproduced by kind permission of the Syndics of Cambridge University Library.*

Overjoyed, and short on time to prepare, Charles set about buying instruments and a new pistol and rifle, getting packed, and meeting with Captain Robert FitzRoy, who would command the ship. The ship was a bit of a shock. The HMS *Beagle* was *small* — only ninety feet long and twenty-four feet wide at the most. It had just two tiny cabins (Figure 1.3). Charles, at six feet tall, had to stoop to enter what for years would be his quarters, and he would share the space with a large chart table, a nineteen-year-old officer,

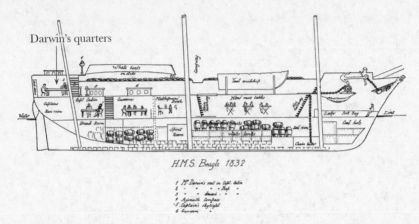

FIGURE 1.3 *The HMS Beagle and Darwin's quarters.*
Based on a drawing by shipmate Philip King, with whom Darwin shared his
quarters. *From* Journal of Researches into Geology and Natural History of the Various
Countries Visited by H.M.S. Beagle *by Charles Darwin (facsimile edition of 1839
First Edition, Hafner Publishing Company, New York, 1952).*

and a fourteen-year-old midshipman. Charles was to sleep in a hammock
that was slung over the chart table and just two feet below a skylight.

Charles made his rounds saying goodbye to friends and family, and
seeking out last minute advice from naturalists. Henslow, who had made the
voyage possible, gave Charles a parting gift — a copy of Humboldt's *Personal
Narrative* — and suggested that he take along Charles Lyell's new *Principles of
Geology* to help him decipher the landscapes he would see. These books, and
a copy of the Bible, would be the young divinity student's close companions.

Saying goodbye to his father was the hardest. Charles would be away for so
long. (Little did he know how long it would be. The voyage was supposed to
take just two years, but it ended up lasting almost five years.) There was also
the very real risk that he might not return. Charles tried to put these thoughts
out of his mind as Ras saw him off from Plymouth.

The *Beagle* made what turned out to be three attempts to start its voyage.
The initial launch, on December 10, 1831, was thwarted when the ship hit a
strong gale. Captain FitzRoy then set out again at low tide on December 21,
only to run the ship aground, and even though he was able to free the ship,
another gale turned it back to shore. These were not good omens for the long

voyage ahead. Finally, on December 27, 1831, the *Beagle* left England and set out for the Canary Islands and South America.

Voyaging

It did not take very long for Charles to be miserable. As the *Beagle* tossed in the waves, Charles tossed everything he tried to eat. Retreating to his hammock, he wondered if the voyage was a huge mistake. He pulled out Humboldt for a little encouragement and tried to look forward to setting foot on land again.

After ten days of torment, the ship finally reached Tenerife in the Canary Islands. From the ship, Charles finally saw the great mountain of which Humboldt wrote, but his excitement was short-lived. The *Beagle* was to be quarantined for fear of its sailors spreading the cholera that had erupted in England just before the *Beagle* set sail. Captain FitzRoy wasn't going to wait, he ordered the sails up, and without anyone stepping on the shores of Tenerife, the ship left for the Cape Verde Islands further south off the west coast of Africa.

St. Jago (San Thiago) would provide Charles relief from his desperate seasickness. Although a volcanic island, Charles was thrilled to see for the first time a tropical landscape. The birds, palms, and massive baobab trees made deep impressions. He also enjoyed the island geology and was intrigued by a band of shells and corals that lay about thirty feet above sea level. Charles, fresh from Sedgwick's training and reading Lyell, began to wonder, had the sea level fallen or the island risen? He would ponder the same questions many more times in the coming years.

After a few weeks of respite, it was back on board the *Beagle* for the crossing to Brazil. The nausea returned and was compounded by the oppressive heat as the ship crossed the equator. Charles was laid up in his cabin, feeling as though he was being "stewed in . . . warm melted butter."

Having retched his way across the Atlantic, he was understandably eager to get off the boat as soon as it made landfall at Bahia, on the coast of Brazil. Charles headed for the forest, and it did not disappoint. His senses were flooded with the colors of the flowers, fruits, and insects, the scents of the plants and trees, and the chorus of all the animal sounds. He wrote to

Henslow, "I formerly admired Humboldt, I now almost adore him; he alone gives any notion of the feelings that are raised in the mind upon entering the Tropics." Charles began to collect everything he could.

After a couple of weeks in Bahia, the *Beagle* sailed on down the coast of Brazil to Rio de Janeiro, from where Charles would again venture out. This was to be a repeating pattern of the voyage. The *Beagle* would sail from port to port conducting its surveys and mapmaking, while Charles would head inland to collect. Captain FitzRoy was obsessive about his work, which bought Charles a lot of time for his land excursions.

Onboard ship, there was also a routine. In a letter to his sister Charles explained:

> We breakfast at eight o'clock. The invariable maxim is to throw away all politeness — that is, never to wait for each other, and bolt off the minute one has done eating, &c. At sea, when the weather is calm, I work at marine animals, with which the whole ocean abounds. If there is any sea up I am either sick or contrive to read some voyage or travels. At one we dine. You shore-going people are lamentably mistaken about the matter of living on board. We have never yet (nor shall we) dined off salt meat . . . At five we have tea.

Charles himself would procure a good portion of the meat his shipmates would consume in the course of the voyage. He was, thanks to his boyhood, a good shot. His skills put him in good stead with the crew of the *Beagle*.

Charles also used the port stops to find a vessel heading home that could carry his specimens. Eight months into the voyage, he shipped his first box to Henslow, for safe-keeping.

Gauchos and Bones

Forays into the interior of a region required some local knowledge, and Charles was usually able to find various native characters willing to accompany him. At Bahia Blanca, a settlement on the coast of Argentina at the edge of the great Patagonian plains or "Pampas," he found himself in the company of "gauchos," the local type of cowboy that Charles found "by far the most savage picturesque group [he] had ever beheld." Amused by their colorful dress and ponchos, Charles also noted the sabers and muskets they carried. They were in constant conflict with local tribes, but as they were also "well known as perfect riders" and knew where to find the few sources of fresh water, Charles and the officers of the *Beagle* accepted their assistance. This included introduction to the local cuisine of rhea eggs (a flightless bird Charles referred to as an "ostrich") and armadillos, which Charles declared "taste & look like a duck."

While exploring the coast just a bit further south near Punta Alta, Charles found some rocks containing shells and the bones of large animals. He used his pick axe to free what he guessed were parts of a "rhinoceros." The next day he found a large head embedded in soft rock and spent so many hours removing it he returned to the ship after dark. Two weeks later, he found a jaw bone and a tooth that he thought belonged to the *Megatherium*, or giant ground sloth. He was not sure of what he had, but crated the bones ("cargoes of apparent rubbish," Captain FitzRoy teased) for shipment so that the experts back in England could decipher their identities.

Eventually, it would be determined that Charles had found remnants of several species, including: a giant armadillo-like creature called a *Glyptodon*; *Toxodon*, an extinct relative of the capybara; and three types of ground sloth — *Megatherium*, *Mylodon* (Figure 1.4), and *Glossotherium*.

It would be a long wait before Charles would even hear that his fossils arrived safely. Safe passage of either parcels or passengers was far from guaranteed in those days, as Charles was about to find out firsthand.

Land of Savages

The *Beagle* continued to sail south along the eastern coast of South America, making its way toward Tierra del Fuego and Charles' first encounter with humans in their most primitive state.

He had been looking forward to the experience. On the previous voyage Captain FitzRoy had taken several native Fuegians back to England where they were clothed and taught in the British fashion. Now, on the return, three of these former "savages" were to be returned to their people in the hope that they might spread some civilization to this part of the world.

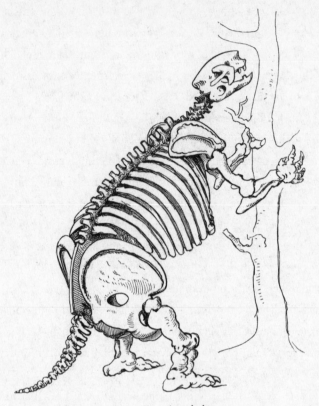

FIGURE 1.4 A *mylodon.*
Drawing of a giant ground sloth. *From* A Naturalist's Voyage Around the World:
The Voyage of the H.M.S. Beagle *by Charles Darwin (D. Appleton and Co.,*
New York, 1890).

Charles rowed ashore with Captain FitzRoy to meet the native Fuegians. He was shocked by their appearance and behavior, and could not believe that the three missionaries they were about to deliver were only recently just as untamed (Figure 1.5). The contrast set in motion much thought about the differences, or lack thereof, between savage and civilized humans.

The *Beagle* forged on to round the notorious Cape Horn. Hugging the coast, the going was rough and the ship often tucked into coves to escape the weather. Charles tried to enjoy the scenery and wildlife, but two weeks of the cold, wind, and waves took their toll. He noted in his diary, "I have scarcely for an hour been quite free from seasickness: How long the bad weather may

FIGURE 1.5 *A Fuegian.*
Drawing of a native of Tierra del Fuego at Portrait Cove. *From* Journal of
Researches into Geology and Natural History of the Various Countries
Visited by H.M.S. Beagle *by Charles Darwin (facsimile edition of 1839
First Edition, Hafner Publishing Company, New York, 1952).*

last, I know not; but my spirits, temper, and stomach, I am well assured, will not hold out much longer."

But the weather worsened, the *Beagle* lost track of its position, and took a pounding. A great wave struck the ship, and the crew had to cut away one of the whale boats. The sea poured onto the decks and started to fill the cabins. Fortunately, once the portholes were opened, the little ship righted herself and water was drained away. One more wave, Charles knew, and that would have been the end. It was the worst gale Captain FitzRoy had ever experienced. Terrified, Charles wrote in his diary, "May Providence keep the *Beagle* out of them."

The *Beagle* inched up the coast, the crew looking for a place to establish a settlement for their Fuegian missionaries. Afterwards, they entered the Beagle Channel. The scenery was magnificent. Glaciers extended from the mountains down to the water, where they calved small icebergs — creating the effect of a small "arctic ocean." But the tranquility of the scene was deceiving. While dining on shore near a glacier, a large ice mass broke off and hit the water, sending a great wave toward the landing party boats on the shore. Charles acted quickly as he and several sailors grabbed the boat lines before the waves could steal the boats away. Had they lost the boats, they would have been in a dire situation, stranded with no supplies in hostile country.

Captain FitzRoy was impressed by Charles' actions and the next day named a large body of water "Darwin's Sound" after "my messmate, who so willingly encountered the discomfort and risk of a long cruise in a small loaded boat." Captain FitzRoy also named a mountain peak in Charles' honor. Charles certainly appreciated the captain's gestures. It was flattering, as a then just twenty-four-year-old geologist, to have features named for him, even if they were at the remote tip of the continent.

But, as the second year of the voyage unfolded and Charles continued his expeditions and collecting, he was increasingly concerned with how his efforts were being received back home in England. He had shipped barrels of specimens and more fossils, including a nearly complete *Megatherium*. With the long wait between the sending of a shipment or letter, and the receipt of a reply, Charles was worried. Had his shipments even reached Henslow? Was he collecting anything of interest? His constant seasickness and bouts of homesickness were also wearing on him. He confessed in one of his letters to Henslow his anxiety about the length of the voyage, "I know not, how I shall be able to endure it."

When the *Beagle* arrived in the Falkland Islands in March 1834, there was mail waiting, and Charles finally got his answer. In a letter he had composed six months earlier on August 31, 1833, Henslow reported that Charles' fossil *Megatherium* "turned out to be most interesting" and had been shown at the meeting of the British Association for the Advancement of Science that summer. The mentor then gently encouraged his pupil, "If you propose returning before the whole period of the voyage expires, don't make up your mind in a hurry . . . I suspect you will always find something to keep up your courage." Then he added, "Send home every scrap of *Megatherium* skull you can set your eyes upon — *all fossils* . . . I foresee that your minute insects will nearly all turn out new."

Henslow's news and encouragement were just what Charles needed. He returned to his geology and collecting with zeal, and he looked forward to the next sights on the voyage — the west coast of South America and the Andes.

Shaky Ground

The price for every new adventure was yet another confrontation with the sea. To get to the west coast of the continent, the *Beagle* sailed through the Strait of Magellan — a "shortcut" that avoided the treacherous Cape Horn (Figure 1.6). But, in late May and early June, nearly winter in the southern hemisphere, it was no leisurely cruise. Charles watched the ice form on his skylight while he clung to his hammock.

There were many reminders of the peril of each leg of the voyage. On the way north to Chile, a shipmate died and was buried in a solemn service at sea. Later, as the *Beagle* scouted islands off the Chilean coast, the crew caught a glimpse of a man waving a shirt and a party was sent ashore to investigate. They found five American crew members who had run away from a whaling ship in a small boat and wrecked before they could reach the mainland. Charles saw the men were in desperate shape, having survived for over a year on nothing but shellfish and seal meat.

FIGURE 1.6 *HMS Beagle in the Strait of Magellan.*
Drawing from A Naturalist's Voyage Around the World: The Voyage of the
H.M.S. Beagle *by Charles Darwin (D. Appleton and Co., New York, 1890).*

Back on the mainland, Charles enjoyed more geological excursions. He found beds of modern shells at an elevation of thirteen hundred feet, and in the Andes he found fossil shells at thirteen thousand feet. How could the marine creatures wind up so high above the sea?

In a forest near Valdivia, he got part of his answer. While resting on a morning walk, he felt the earth tremble, and then shake so violently he could not stand. He went back into town and found chaos. Houses were tilting and the citizens were in shock.

As the *Beagle* sailed north, the devastation was everywhere. The city of Concepción was pummeled to rubble. The inhabitants described the earthquake as the worst ever; it had also triggered a tsunami and widespread fires. Many people were still buried.

At the shore, Charles observed that *the mussel beds were now positioned several feet above the water.* The new position of the mussels was it — proof that the land had been uplifted. The great mountains were built in small steps, just as Lyell had written, and now Charles was an eyewitness to the process.

On a slope in the Andes, he found even more stunning testimony. At seven thousand feet there was a grove of fossilized trees. How could trees be sitting this high up, embedded in sandstone? Charles deciphered the geological explanation for this astonishing sight:

> I saw the spot where a cluster of fine trees had once waved their branches on the shores of the Atlantic, when that ocean (now driven back 700 miles) approached the base of the Andes . . . upright trees, had subsequently been let down to the depths of the ocean. There it was covered by sedimentary matter . . . but again the subterranean forces exerted their power, and I now beheld the bed of that sea forming a chain of mountains more than seven thousand feet in altitude.

These mountains were bearing trees that had once been buried in the sea bed.

Land sinking, mountains rising — Charles began to think about everything he saw in a dynamic geological perspective. On the island of San Lorenzo off the coast of Peru, he examined the shell beds which rose above the level of the sea. In a terrace at the eighty-five-foot level, he was very curious to find, together with the shells, cotton thread, plaited rush (braided seagrass), and the head of a stalk of Indian corn — signs of earlier human inhabitants. Charles deduced that the island had risen eighty-five feet since humans last lived there.

Geology dominated Charles' thoughts. Off the coast of Peru he began to think about the Pacific islands he was about to visit. One of the *Beagle's* assignments was to take measurements around the picturesque coral islands and to see whether the rings of coral that encircled them sat upon the rims of rising volcanic craters, as then believed. Though he had not yet seen a coral island with his own eyes, Charles rethought the situation and came to the opposite conclusion. What if the mountains were actually sinking? Then the corals, which required the light of shallow water, would encircle and grow upward around the sinking masses. If so, the beautiful rings of coral atolls were not sitting on crater rims, but encircling sinking land forms. This was the first theory he could call his own.

Charles wrote to Henslow that he was looking forward to his next stop for two reasons: it would bring him that much nearer to England, and it would give him a chance to see an active volcano. But this time it would be the animals and not the landscape that set his mind in motion. The *Beagle* sailed for the Galapagos Islands some six hundred miles off the coast (Figure 1.7).

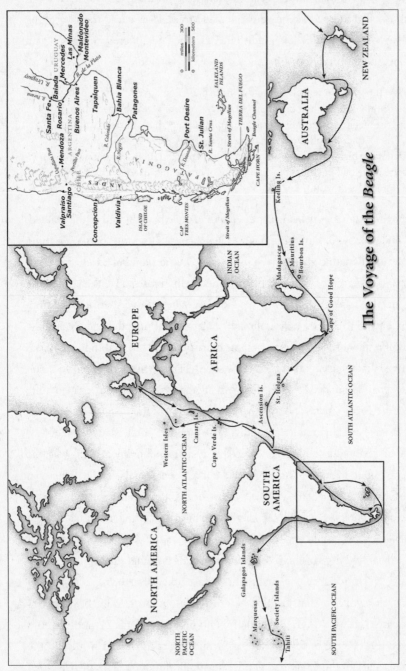

FIGURE 1.7 *Map of the Voyage of the HMS Beagle, 1831–1836. Drawn by Leanne Olds.*

Reptile Paradise

Charles arrived in the Galapagos on September 15, 1835, well into the fourth year of the voyage. One might think that with these islands now inextricably linked with Darwin's name, they were the young naturalist's Eden. Far from it. In his diary of the first days there he wrote:

> The stunted trees show little signs of life. The black rocks heated by the rays of the vertical sun like a stove, give to the air a close & sultry feeling. The plants also smell unpleasantly. The country was compared to what we might imagine the uncultivated parts of the infernal regions to be.

But he did find a bay swimming with fish, sharks, and turtles, and described the islands as:

> . . . paradise for the whole family of Reptiles . . . The black lava rocks on the beach are frequented by large (2–3ft.) most disgusting, clumsy lizards . . . They assuredly well become the land they inhabit.

On a stroll, Charles encountered:

> . . . two very large Tortoises (circumference of shell about 7ft.). One was eating a Cactus & then quietly walked away . . . They were so heavy I could scarcely lift them off the ground. Surrounded by the black lava, the leafless shrubs & large Cacti, they appeared most old-fashioned antediluvian animals; or rather inhabitants of some other planet (Figure 1.8).

He found great numbers of tortoises near the freshwater springs and was amused by the lines of animals marching to and fro.

On James Island, Charles collected all of the animals and plants he could. He was curious to decipher whether the plants were the same as those on the South American continent or if they were peculiar to the Galapagos Islands. He also paid attention to the birds. The species of mockingbird on James Island looked different from those on other islands. Moving from island to island, the primary challenge was collection; identification would come later.

Charles did resolve a mystery of the marine iguanas and what they ate. An earlier visitor had concluded that the lizards went out to sea to fish. But Charles opened the stomachs of several of the animals and found that they

FIGURE 1.8 A *Galapagos tortoise.*
Drawing from A Naturalist's Voyage Around the World: The Voyage of the H.M.S.
Beagle *by Charles Darwin (D. Appleton and Co., New York, 1890).*

were packed with the seaweed that grew in thin layers on the submerged rock. Though hideous to his eye, Charles did admire the animals' great swimming ability and diving endurance. He noted that he believed these habits were unique to marine iguanas — not seen in any other lizard — and strikingly different than those of the Galapagos Islands' land iguanas.

After five weeks of hiking across sand that reached 137°F, Charles and the *Beagle* departed for points west.

Centers of Creation and the Mystery of Mysteries

Charles, for once, enjoyed the long sail through tropical seas to Tahiti. One of his cabin mates, Midshipman King, would later recall the pleasure Charles took, "in pointing out to me as a youngster the delights of the tropical nights, with their balmy breezes eddying out of the sails above us, and the sea lighted up by the passage of the ship through the never-ending streams of phosphorescent animalculae."

The *Beagle* went on to New Zealand, Australia, and then to the Cocos Islands, where Charles saw his first coral atolls, with their reefs encircling gorgeous blue lagoons. Wading among the corals, he immersed himself in the wonders of the reef and confirmed his suspicions about how such beautiful structures were built.

The *Beagle* then sailed on toward Africa and reached it at the end of May, 1836. At the Cape of Good Hope, Charles went ashore with the captain to call upon the great astronomer, Sir John Herschel, whose book Charles had read while studying at Cambridge. Herschel was keenly interested in geology and was also a close follower of and correspondent with Lyell. But Herschel thought Lyell had missed the mark with his second volume of the *Principles of Geology*.

The second volume, a copy of which Charles received at sea in a shipment from home, focused on questions surrounding the appearance of species. Lyell refuted the then circulating ideas of evolution and proposed instead that species were created "in succession at such times and in such places as to enable them to multiply and endure for an appointed period, and occupy an appointed space on the globe." Lyell proposed that the congregations of species peculiar to any area marked "centres or foci of creation . . . as if there were favorite points where the creative energy has been in greater action than others . . ." Lyell was adhering to the conventional view of species as unchanging — with each kind specially created — while explaining the succession of species he knew so well from the fossil record as a succession of creations.

But Herschel thought otherwise. If landscapes evolved, as Lyell had amply demonstrated, why not their inhabitants? Herschel saw a connection to the "mystery of mysteries" — the origin of new species. Whether Herschel fully disclosed his thoughts on the matter is not clear. But it is clear that on the journey home, and thereafter, the mystery of mysteries gripped Charles.

Charles had much to look forward to on the way back to England. Henslow had collected and published ten of his letters in a pamphlet, and

Charles' sister wrote that his name was gaining much attention in England. Charles began to plot his return and to prioritize his work. Surrounded by reams of geological, zoological, and botanical notes, he began organizing them for publication. The final leg of the voyage, thanks to Captain FitzRoy's obsessive chartmaking, was going to be longer than expected. Instead of heading up the west coast of Africa to Europe, the ship headed back to Brazil for one last check of measurements. Charles, who never overcame his seasickness, wrote home: "I loathe, I abhor the sea."

But Charles reconciled to make good use of the extra time. He began to gather and to flesh out his ornithological notes, and returned to the puzzle of the Galapagos birds. He concluded that the mockingbirds of the Galapagos were closely allied in appearance to those of Chile. But there was more to the story. Charles wrote in his notes:

> I have specimens from four of the larger islands; the specimens from Chatham & Albemarle Isd. appear to be the same, but the other two different. *In each Isd. each kind is exclusively found* [emph. added]; habits of all are indistinguishable. When I recollect, the fact that from the form of the body, shape of scales & general size, the Spaniards can at once pronounce from which Isd. any tortoise may have been brought: — when I see these Islands in sight of each other and possessed of but a scanty stock of animals, tenanted by these birds but slightly differing in structure filling the same place in Nature, I must suspect they are only varieties. The only fact of a similar kind of which I am aware is the constant asserted difference between the wolf-like fox of East & West Falkland Isds. — *If there is the slightest foundation for these remarks, the zoology of Archipelagoes — will be well worth examining; for such facts would undermine the stability of species.*

By the end of the voyage Charles was already pondering that mystery of mysteries in a fresh light.

The Mariner Returns

It was a joyous and triumphant homecoming.

For five years Charles had been away from his friends, family, and mentors. His sisters were so relieved to see him home safe. And The Doctor, well, he was very proud. His son had left a directionless bug-catcher, but had returned to the toasts of the cream of British scientific circles. Charles was most anxious to see Henslow again — and to get his advice on what to do with the specimens he had collected.

The great Lyell wanted to meet him, and Charles soon was invited to a dinner at the London home of his geological hero. Lyell was transfixed by Charles' tale of the Chilean earthquake, and introduced him to the people who could help with the scientific analysis of his collections. The fossils, the birds, the plants, and even the iguanas found eager takers.

Charles was pondering writing a book about his long voyage. He loaned his Wedgwood cousins his diaries to read and was very encouraged by his cousins' reactions to his adventures. Captain FitzRoy developed plans for a three-volume "narrative" of the *Beagle's* voyages, written by himself, another previous captain of the ship, and Charles.

As Charles set to work writing his account of the *Beagle* voyages, the experts were poring over his collections. Ornithologist John Gould, an accomplished naturalist and illustrator, quickly perceived that Charles' Galapagos birds were closely related. What Charles thought were "gross-beaks" and "blackbirds" were actually finches. In just a few days' examination, Gould had identified twelve (later revised to thirteen) species of ground finches — all entirely new species (Figure 1.9). And the "varieties" of Galapagos mockingbirds? They included three distinct species. They were related to those in Chile, as Charles surmised, but they were not identical to them.

So here was the crucial puzzle. How could Charles explain all of these new species, each specific to an island? The conditions on each island were not significantly different, so if each bird had been created to suit each island, then why were the birds different? It was inescapable. The original birds that immigrated to the islands had changed somehow and produced new species.

Charles knew this was difficult to explain, even more difficult to persuade others of, and as it violated the doctrine of the immutability of species and challenged creation-based explanations, very dangerous territory. He was torn. He was eager for recognition and to climb the ranks of the scientific elite, but he knew that the "transmutation" of species was professional suicide. Neither his new boosters in the scientific community nor his Cambridge mentors would stand for such heresy.

1. Geospiza magnirostris.
2. Geospiza fortis.
3. Geospiza parvula.
4. Certhidea olivacea.

FIGURE 1.9 *Galapagos finches.*
Drawing from A Naturalist's Voyage Around the World: The Voyage of the
H.M.S. Beagle *by Charles Darwin (D. Appleton and Co., New York, 1890).*

He worked furiously on his journal and tried to finesse the issues raised by
the Galapagos animals:

> It never occurred to me, that the productions of islands only a few miles apart,
> and placed under the same physical conditions, would be dissimilar. I therefore
> did not attempt to make a series of specimens from the separate islands. It is the
> fate of every voyager when he has just discovered what object in any place is
> more particularly worth of his attention, to be hurried from it . . . It is clear, that
> if several islands have their peculiar species of the same genera, when these are
> placed together, they will have a wide range of character. *But there is not space
> in this work, to enter on this curious subject.*

And so began the dodging game he would play for the next twenty years.
When he wrote these lines, he was already convinced that species change, but
he did not tip his hand. Indeed, when a young Alfred Russel Wallace read this
passage he saw the mystery of mysteries as a still-open question that the great
Darwin had overlooked, and this spurred Wallace to make his own voyages (as
will be seen in Chapter 2).

Charles finished his *Journal of Researchers into the Geology and Natural
History of the Various Countries Visited by the H.M.S. Beagle Under the*

Command of Captain FitzRoy, R.N. from 1832 to 1836 (what became known as *The Voyage of the Beagle*) in seven months. (It would not appear for two years due to Captain FitzRoy's delays in finishing his part.)

Secret Notebooks and a Species Theory

Publicly, Charles would only go so far. Privately, he threw himself into the study of the transmutation of species.

He recalled the "ostriches" of South America. Early in the voyage, he had heard of a second, smaller form of the bird that lived in southern parts of Patagonia beyond the Rio Negro. The *Petise* form was rare, and Charles wanted very badly to collect one but had no luck, as the bird was difficult to spot and very wary. While dining one night on what he casually thought was a juvenile ostrich, it dawned on him that he was actually consuming the elusive *Petise* species. Panicked, he rescued some parts that had not been cooked or eaten. Years later, after Charles returned, John Gould named the reassembled bird *Rhea darwinii*.

Now Charles began puzzling over the fact that the large and small rheas overlapped in their territories near the Rio Negro. Unlike the Galapagos birds, there was no physical boundary to separate them. The two species made Charles note, "One is urged to look to common parent?" Charles opened up a new notebook (called simply "B") and jotted his thoughts down as they streamed out. On page fifteen, he recalled the animals of Australia and scrawled:

> Countries longest separated greatest differences — if separated from immens [sic] ages possibly two distinct type, but each having its representatives — as in Australia. This presupposes time when no Mammalia existed; Australia Mamm. were produced from propagation from different set, as the rest of the world.

On page twenty:

> We may look at Megatherium, armadillos, and sloths as all offsprings of some still older type some of the branches dying out . . .

On page twenty-one:

> Organized beings represent a tree *irregularly branched* some branches far more branched — Hence Genera. As many terminal buds dying as new ones generated . . .

On page thirty-five:

> If we grant similarity of animals in one country owing to springing from one
> branch . . .

And then on page thirty-six, following the declaration "I think," he drew a little diagram that represented a new system of Natural History, a tree of life
with ancestors at the bottom and their descendants at the top (Figure 1.10).

His jottings raced from topic to topic in zoology, geology, and anthropology.
Each entry was a fragment of a much larger picture that was slowly taking form.

Life was a tree, with the branches and twigs connecting species, like relatives in a family pedigree. But what made the tree branch? Why were new
forms arising and others dying out?

Through the next year, he read all sorts of books for hints toward the questions burning in his mind. On September 28, 1838, he opened Thomas Malthus'
Essay on the Principle of Populations. Malthus proposed that there were
checks — disease, famine, and death — upon the growth of populations that
prevented them from increasing at an exponential rate. Malthus explained that
there was great overproduction of offspring in nature because of these checks.
And what would sort out the survivors from the others? It was clear to Charles,
the stronger, better adapted ones.

The result, Charles realized, would be the formation of new species.

His "species theory" was thus born, and would grow and develop over the
next few years. He quickly made an analogy between the role of nature in
shaping her species and the role of humans in shaping their animal breeds: "It
is a beautiful part of my theory that domesticated races of organisms are made
by precisely the same means as species — but latter far more perfectly & infi-
nitely slower." That natural process was to be called "natural selection."

Charles was also recalibrating the clock of life. Influenced by astronomer
Herschel, who suggested that "the days of Creation" may correspond to "many
thousand millions of years," Charles felt confident that the earth and life were
much older than geologists had grasped.

But even as Charles' certainty increased, all of these ideas remained private.
He kept his notebooks and his ideas secret because they were heretical to the

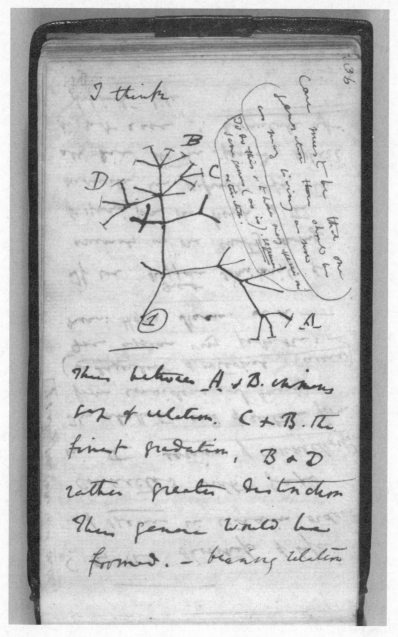

FIGURE 1.10 *The tree of life.*
Page from notebook "B," where Darwin recorded his idea that life is connected
like the branches of a tree, with ancestors at the bottom. *Reproduced by kind
permission of the Syndics of Cambridge University Library.*

doctrine of special creation, which was held sacred by those in power in Britain — the government, the Church, and fellow scientists, including those who trained and supported Charles. It would be professional suicide to go public. He was young and not established. Charles was getting plenty of recognition without having to make public such radical thoughts.

In 1839, *The Voyage of the Beagle* appeared to great acclaim. Charles was becoming famous. One day, a letter arrived from Potsdam. It was from the great Humboldt himself. Gushing with praise, Humboldt declared that his influence on Charles was "the greatest success that my humble work could bring."

Charles was thrilled and moved. He answered his hero thanking him for the great pleasure his letter had given. Charles wrote, "That the author of those passages in the Personal Narrative, which I have read over and over again, & have copied out, that they might ever be present in my mind, should have so honoured me, is a gratification of a kind, which can but seldom happen to anyone." It was too much of a risk to gamble his soaring reputation on his species theory.

The Dear Old Philosopher

There was neither the time nor the inclination to return to his divinity studies. Charles was fully consumed by his science and felt that if he did not work very hard on the fruits of his voyage in the first few years home, he would be overwhelmed. While his heart was no longer there for his parsonage, he was eager to settle down and start a family. Just before he turned thirty, he married his first cousin Emma Wedgwood.

They had known each other all of their lives. Charles confided to Emma where his thoughts were leading. Emma, a devout Christian, worried that Charles' heresies might preclude their eternal life together. It was a delicate balance. Emma knew that Charles was working on great ideas, but Charles was very mindful of Emma's concerns. Charles had yet another reason to keep his theory private.

In 1842, Charles distilled his notes and several years' thinking into a thirty-five-page sketch of his species theory. Two years later, he expanded the sketch into a work of 230 pages. The table of contents is strikingly similar to that of *The Origin of Species*, which would not appear for seventeen years, in 1859. Many of the well-known arguments and prose from his great book appear in both of these early manuscripts.

But Charles still thought it was unwise to publish. He would, in time, share the long essays only with a few trusted intimates — Lyell; botanist Joseph Hooker; biologist Thomas Huxley; and Emma. On July 5, 1844 Charles wrote his wife a note:

> I have just finished my sketch of my species theory. If, as I believe, my theory in time [will] be accepted even by one competent judge, it will be a considerable step for science . . . I therefore write this in case of my sudden death, as my most solemn and last request . . . that you will devote £400 to its publication, and further will yourself, or through Hensleigh [Emma's brother], take trouble in promoting it.

More evidence was what Charles decided would be needed to bolster his theory. He then went to work on all sorts of topics in botany, zoology, and geology. Independently wealthy, thanks to his father and his father-in-law, and ensconced at his manor in the village of Down, his life revolved around his work, Emma, and their ten children (seven of whom lived to adulthood). He was a doting father and often regaled his children with tales of his adventures and stories about his shipmates on the *Beagle*.

Charles' demeanor as a father and a husband was the same as his disposition at sea. His *Beagle* shipmates could not recall ever seeing Charles out of temper or hearing him say an unkind word about or to anyone in the course of their long journey. It was in admiration of these qualities and his abilities that Charles' shipmates had given him perhaps his most apt nickname, "the dear old Philosopher."

The Voyage of the Beagle would in turn inspire another wave of naturalists, for whom Charles would play the same role as Humboldt had in his day. And a letter from one of this new group of voyagers would finally, after twenty years, make him break his silence about the origin of species and finally publish his greatest work.

CHAPTER QUESTIONS

1. What experiences of his youth helped to prepare Darwin for the voyage on the *Beagle*?

2. What geological phenomena and formations did Darwin witness? How did these shape his thinking about the age of the earth or how life changed?

3. What zoological evidence led Darwin to think that species evolved?

4. What were Thomas Malthus' ideas, and how did Darwin react to them?

5. Why did Darwin delay publishing his species theory?

For more on this story, go to the *Into The Jungle* companion website at www.aw-bc.com/carroll.

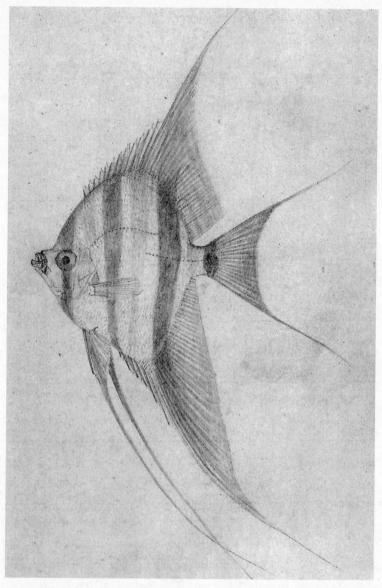

FIGURE 2.1 *Sketch salvaged from fire and shipwreck of the Helen.*
This drawing of an Amazonian angelfish was one of the few sketches Wallace
managed to save out of all of his notes and specimens on his doomed voyage home. It
displays one of the important talents for naturalists before the age of photography —
that of being a good artist. *Drawing from the autobiography of Alfred Russel Wallace,
My Life (New York: Dodd, Mead, and Co., 1905).*

Drawing the Line between Monkeys and Kangaroos

"All truths are easy to understand once they are discovered; the point is to discover them."

—Galileo Galilei

It was time to pack up and go home.

Alfred Wallace was two thousand miles upriver from the Atlantic Ocean, on the Rio dos Uaupés tributary of the Amazon — further than any European had ever gone. Since arriving in May 1848, he had spent nearly four years exploring and collecting, but had been laid up the last three months with yellow fever. He was too exhausted to go on. His younger brother, Herbert, who accompanied him up to the Rio Negro, had long before turned back. Unbeknownst to Wallace, Herbert was stricken with yellow fever and died before he could board a boat to England.

Wallace had accumulated a large menagerie of live animals — monkeys, macaws, parrots, and a toucan — that he hoped to take all the way to the London Zoo. Their upkeep was killing him. Besides the animals, he also had a couple of year's worth of specimens, both with him and stored downriver, that he had not yet been able to ship to England for sale.

Wallace began to dream of green fields, neat gardens, bread and butter, and other comforts of home. On July 12, 1852, he boarded the brig *Helen* with thirty-four live animals, many boxes of specimens and notes, and set sail for England.

"I'm afraid the ship's on fire; come and see what you think of it." Just after breakfast, three weeks out of port and somewhere east of Bermuda, the captain of the *Helen* was concerned enough to visit Wallace in his cabin. And rightfully so — smoke was pouring out of the hold.

The crew tried, but could not douse the smoldering blaze. The captain ordered down the lifeboats. Wallace, still weak from his bout with yellow fever, looked on as if the scene was a feverish dream. It wasn't. He reentered his hot, smoky cabin and salvaged a small tin box and threw in some drawings (see Figure 2.1), some notes, and a diary. He grabbed a line to lower himself into a lifeboat, slipped, and seared his hands on the rope. His pain was compounded when his injured hands hit the salt water. Once in the lifeboat, he discovered it was leaking.

Wallace watched his animals perish, and then the *Helen*, along with all of his specimens.

And so there he was, lying on his back in a leaky lifeboat in the middle of the Atlantic. His loss did not yet fully dawn upon him. He was optimistic about being rescued and enjoyed the dolphins frolicking around him. But the wind changed, and day after day passed in the open boat. Wallace was blistered with sunburn, parched with thirst, soaked by sea spray, exhausted by constantly bailing water, and near starvation. At last, on the tenth day, Wallace and the other passengers and crew were picked up.

That first night aboard his rescue ship, the *Jordeson*, Wallace could not sleep:

> It was now, when the danger appeared past, that I began to feel fully the greatness of my loss . . . How many weary days and weeks had I passed, upheld by the fond hope of bringing home many new and beautiful forms from those wild regions; every one of which would be endeared to me by the recollections they would call up, which should prove that I had not wasted the advantages I had enjoyed and would give me occupation and amusement for many years to come! And now everything was gone, and I had not one specimen to illustrate the unknown lands I had trod . . .

Actually, the danger was not yet past. The *Jordeson* was twice hit by heavy storms. The first tore the sails and sent a wave crashing through Wallace's skylight, soaking him while he slept. The second hit in the English Channel when they were almost home. That storm sank many ships and put four feet of water in the *Jordeson*'s hold.

Wallace swore to himself "fifty times" on the voyage home: "If I once reached England, never to trust myself more on the ocean." If he had held to that promise, his story would end here and few would have ever heard of Alfred Wallace again.

He broke his vow within days.

Where to Next?

Thank goodness for his choice of Samuel Stevens, the agent who sold what Wallace had managed to ship to England before his calamity. Stevens met Wallace in London, bought him a new suit, and Stevens' mother fed Wallace until his strength returned. Moreover, Stevens had the foresight to insure Wallace's collections for two hundred pounds. It was not as much money as Wallace had hoped to gain by selling his Amazonian treasure, but it was enough to keep him from begging.

The loss of his specimens, rather than deterring Wallace from further adventures, only stoked his determination. His voyage was unfinished. His lust for exploration and collecting was not satisfied, nor was his obsession with questions about how new species formed. Now thirty, he was, unlike Darwin, not yet ready to settle down.

He began to ponder his next journey. The big question was "where to go?" He had both practical and scientific matters to consider. The fiery young zoologist Thomas Huxley once said, "Science in England does everything — except pay." For working-class, self-made men like Huxley and Wallace, this was painfully true. Wallace had to collect quarry that would fetch good prices. He ruled out a return to the Amazon. His friend and fellow entrepreneur, Henry Walter Bates, was still there and had that covered (see Chapter 3). No, Wallace's next adventure had to be to new territory.

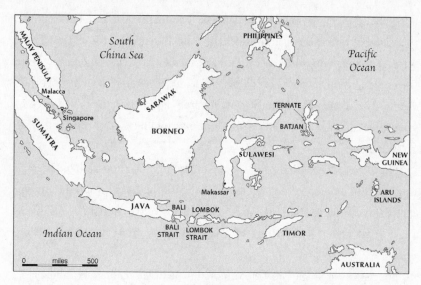

FIGURE 2.2 *The Malay Archipelago.*
Map by Leanne Olds.

Wallace kept thinking about the Malay Archipelago, the vast group of islands between Southeast Asia and Australia (Figure 2.2). Other than on the island of Java, the animals and plants of the archipelago were unexplored. Enough fragments of natural history were emerging from the Dutch settlements there to convince Wallace that the islands offered both rich pickings and good facilities for a traveler. The islands spanned more than four thousand miles from east to west and thirteen hundred miles from north to south, an area almost as large as the entire continent of South America. Many of the islands were volcanic (one, Krakatoa, would nearly vaporize in 1883 in an enormous eruption that altered the planet's climate). Covered in tropical forest, the islands appeared similar, but some held different treasures. Discovering and explaining the differences would put Wallace, literally, on the map.

On the Hunt Again

It was a much longer journey to the Far East than to Brazil. Wallace arrived in Singapore in April 1854, and set out to explore the country. He encountered altogether different treasures, and dangers, than on the Amazon. For example, the island of Singapore was an excellent collecting ground for insects. There were, however, some drawbacks. Wallace noted, "Here and there, too, were tiger pits, carefully covered over with sticks and leaves, and so well concealed, that in several cases I had a narrow escape from falling into them . . . fifteen or twenty feet deep, so it would be almost impossible for a person unassisted to get out of one."

There were tigers roaming about Singapore, and they took on average one resident a day. Wallace occasionally heard the beasts' roars and in typical British understatement, he noted, "It was rather nervous work hunting for insects . . . when one of these savage animals might be lurking close by."

The natives of Singapore were rumored to be just as dangerous as the tigers. Wallace had his mattress on the floor of his bamboo house. He later recalled a friend's admonishment: "There were many bad people about, who might come at night and push their spears up through me from below, so he kindly lent me a sofa to sleep on, which, however, I never used as it is too hot in this country."

Unflustered by such concerns, Wallace followed a daily routine. Up at 5:30 a.m., he started the day with a cold bath and hot coffee. He sorted out the previous day's collection and then set out again into the forest with his gear. He carried a net, a large collecting box hung on a strap over his shoulder, pliers for handling bees and wasps, and two sizes of specimen bottles for large and small insects, attached by strings around his neck and plugged with corks. And, oh yes, on some days he carried a rifle.

To preserve his specimens and skins, he relied on locally brewed arrack. About 70 percent alcohol by volume, arrack was distilled from various fermented fruits, grains, sugarcane, or coconut sap. There was a flourishing trade in the beverage, so Wallace could usually stay well stocked. The only problem was the natives' taste for the drink was very keen, so casks often went missing from his house or field camps. Wallace then developed the defensive strategy of putting dead snakes and lizards into his casks, but even that did not deter many from drinking his supplies.

The natives could not understand why all the animals, birds, insects, and plants were preserved so carefully (using up good arrack!). Wallace told them that people in his country wanted to look at the creatures. This made no sense. Surely there must be better things to see in "Ung-lung" (as one Wanumbai tribesman pronounced *England*). Some tribes Wallace met did seem to enjoy their own version of collecting. The headhunting Dyaks kept bundles of their enemies' heads suspended from the ceilings of their long-houses. How quaint.

Despite some reputation for ferocity, the native tribesmen shared their knowledge of the forest with Wallace and helped him to find what he was after. He stalked the forests' most beautiful and prized treasures — orangutans, monkeys, spectacular birds of paradise, and enormous brilliant butterflies. Wallace mused:

> Nature seems to have taken every precaution that these, her choicest treasures, may not lose value by being too easily obtained. First, we find an open harbourless, inhospitable coast, exposed to the full swell of the Pacific Ocean; next, a rugged and mountainous country, covered with dense forests, offering in its swamps and precipices and serrated ridges an almost impossible barrier to the central regions; and lastly, a race of the most savage and ruthless character . . .

No matter how many years he spent in the forest, the thrill of capturing something new never diminished:

> I had seen sitting on a leaf out of reach, an immense butterfly of a dark colour marked with white and yellow spots . . . I at once saw that it was a female of a new species of Ornithoptera or "bird–winged butterfly," the pride of the Eastern tropics. I was very anxious to get it and to find the male, which in this genus is always of extreme beauty. During the two succeeding months I only saw it once again . . . I had begun to despair of ever getting a specimen . . . till one day . . . I found a beautiful shrub . . . and saw one of these noble insects hovering over it, but it was too quick for me, and flew away. The next day I went again to the same shrub and succeeded in catching a female, and the day after a fine male . . . more than seven inches across the wings, which are velvety black and fiery orange, the latter color replacing the green of the allied species. The beauty and brilliancy of this insect are indescribable . . . on taking it out of my net and opening the glorious wings, my heart began to beat violently, the blood rushed to my head, and I felt much more like fainting than I have done when in apprehension of immediate death. I had a headache the rest of the day . . . (Figure 2.3).

FIGURE 2.3 *The Golden Birdwing butterfly.*
Wallace discovered this form (*Ornithoptera croesus lydius*) on the island of Batjan.
Photograph by Barbara Strnadova.

Thinking Out Loud

That pounding head was doing more than fawning over butterflies. Wallace was thinking constantly of the diversity of species he saw, the varieties of individuals of a species, and about *where he found them.* These were the practical concerns of a paid collector, but also the catalysts to Wallace's transformation into a scientist.

While Darwin was cloistered in his home at Down and still keeping quiet about his species theory, Wallace was thinking out loud, putting his thoughts on paper, and dashing them off to magazines and journals in England. Some of these were short field notes, but others revealed bigger ideas. Wallace was puzzling over some of the same facts and observations as Darwin, and reaching some remarkably similar conclusions. But Wallace had none of the concerns that restrained Darwin. He had a reputation to make, and nothing to lose.

In 1855, while waiting out the wet season in Sarawak, on Borneo, Wallace wove together threads of geology and natural history to propose a new law: *Every species has come into existence coincident both in space and time with a preexisting, closely allied species.*

Wallace thought that species were connected like "a branching tree." He was proposing that new species come from old species as new twigs grow from older branches.

Doesn't sound *too* dangerous, does it? But it was very bold, because Wallace's target was the well-accepted doctrine of special creation — that each species was specially created, in one moment, to fit the land it inhabited. Moreover, Wallace was using some of the very arguments that Darwin had agonized over for almost two decades, but had not yet published.

Wallace embraced the growing picture from the geology of a changing Earth, and the mounting evidence from the fossil record of the obvious changes in life. He simply extrapolated that what was true of the past must be true of the present, and he reasoned "that the present geographical distribution of life upon the earth must be the result of all the previous changes, both of the surface of the earth itself and of its inhabitants." In short, Wallace surmised that the earth and life evolve together. He called this principle his "Sarawak Law." Folks were starting to get used to the idea of the earth changing, but they didn't at all like the idea of life evolving.

Wallace supported his Sarawak Law with all sorts of observations on the distribution of species, especially those on islands. Take the Galapagos, Wallace argued, "which contain little groups of plants and animals peculiar to themselves, but most early allied to those of South America, have not hitherto received any, even a conjectural explanation." (Ouch! Wallace was indirectly criticizing Darwin, who had hitherto dodged the subject of the possible reasons for the diversity in species.) Wallace continued, "They [the Galapagos] must have been first peopled, like other newly-formed islands, by the action of winds and currents, and at a period sufficiently remote to have had the original species die out, and the modified prototypes only remain." (Translation: There are no finch species in South America identical to those on the Galapagos, but there are such close allies that South American finches must have colonized the islands.)

Wallace pointed out that families of birds, butterflies, and various plants are confined to certain regions. He had noticed years before when he was in the Amazon that some species of monkeys were confined to one side of the river. Wallace thought, "They could not be as they are, had no law regulated their creation and dispersion." By *dispersion*, Wallace meant that the extent to which a species could spread out over the land was constrained by features of the land — rivers, mountain ranges, and so forth.

Almost no one read or noticed Wallace's paper in 1855. He heard nothing from England about his Sarawak Law, except for some grumblings that he should focus on collecting and not theorizing.

Wallace kept moving, collecting, and writing. He had much more to see, and to say.

Drawing a Line

Wallace went island-hopping quite often. He made ninety-six journeys totaling about fourteen thousand miles and visited some of the same islands several times over the span of eight years. Often the availability, or unavailability, of a boat determined his path. He had to be flexible. He tried several times to get from Singapore to Makassar on the island of Sulawesi. No luck. But one day in May 1856, he did find a Chinese schooner headed to Bali, which he had no intention of visiting, but he figured he could find a way from there to Lombok and then on to Makassar. This accidental detour would give Wallace the most important discovery of his expedition.

On Bali, Wallace found a similar variety of birds as on the other islands he had visited — a weaver, a woodpecker, a thrush, a starling — nothing too exciting. But then, he later recalled, "crossing over to Lombok, separated from Bali by a strait less than twenty miles wide, I naturally expected to meet with some of these birds again; but during a stay there of three months I never saw one of them." Instead, Wallace found a completely different assortment of birds — white cockatoos, three species of honey-suckers, a loud bird the locals called a "Quaich-Quaich," and a really strange bird called a megapode (meaning "big foot") that used its big feet to make very large mounds for its eggs. None of these groups were known on the western islands of the archipelago — Java, Sumatra, Malaysia, or Borneo. Now here was a puzzle. What constraint prevented the spread of these species from island to island? Surely, birds could cover a twenty-mile strait with little trouble.

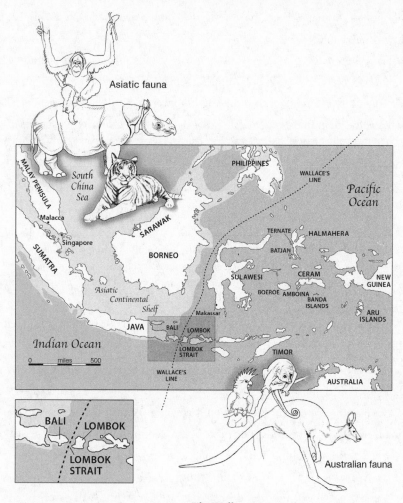

FIGURE 2.4 *The Wallace Line.*
Wallace discovered that the narrow strait between Bali and Lombok marked
a boundary between Asiatic fauna (with tigers, rhinoceri, and orangutans) and
Australia-type fauna (with kangaroos, cuscus, and other marsupials). Bali was once
connected to the Asiatic continental shelf, but not to Lombok. The boundary line
extends throughout the archipelago as shown. *Drawn by Leanne Olds.*

It was as if, Wallace wrote in a letter to Bates, there was some kind of
"boundary line" between Bali and Lombok (Figure 2.4 map). Traveling further east to Flores and Timor, the Aru Islands, and New Guinea, the
changeover in bird life was very clear. All of the families of birds that were

common on Sumatra, Java, and Borneo were absent from Aru, New Guinea, and Australia, and vice versa.

The differences in mammals among the western and eastern islands of Indonesia were just as striking. On the large western islands there were monkeys, tigers, and rhinoceros. But on Aru there were no primates or carnivores, all the native mammals were marsupials — a kangaroo, three or four species of cuscus, and some other small rat-like marsupials.

That line between Bali and Lombok was real, and signified something very profound to Wallace. He put his thoughts to paper again (published in 1857):

> Let us now examine if the theories of modern naturalists will explain the phenomena of the Aru and New Guinea fauna . . . How do we account for the places where they came into existence? Why are not the same species found in the same climates all over the world? The general explanation given is, that as the ancient species became extinct, new ones were created in each country or district, adapted to the physical conditions of that district.

By *created*, Wallace meant "specially created by a Creator." But, Wallace pointed out, a theory of creation would make us expect to find similar animals in countries with similar climates, and dissimilar animals in countries with dissimilar climates. This is not at all what he saw.

Comparing Borneo (in the west) and New Guinea (in the east), Wallace remarked, "It would be difficult to point out two countries more exactly resembling each other in climate and physical features." But, in fact, the regions' birds and mammals were entirely different.

Comparing New Guinea and Australia, Wallace wrote, "We can scarcely find a stronger contrast than in their physical conditions . . . one enjoying perpetual moisture, the other with alternatives of drought." Wallace reasoned,

> If kangaroos are especially adapted to the dry plains and open woods of Australia, there must be some other reason for their introduction into the dense damp forests of New Guinea, and we can hardly imagine that the great variety of monkeys, squirrels, of Insectivores, and Felidae [cats], were created in Borneo because the country was adapted to them, and not one single species given to another country exactly similar and at no great distance.

In the tropical forests of the eastern islands, tree kangaroos occupied the habitat occupied by monkeys in the west.

The reason for the observed differences in animal distributions, Wallace concluded, must be that "some other law has regulated the distribution of existing species." That other law, Wallace suggested, was the Sarawak Law, which he had proposed two years earlier. Again Wallace relied on geology to make his case. He surmised that New Guinea, Australia, and Aru must have been connected at some time in the past and that such a connection would explain why those landmasses shared similar sets of birds and mammals. And the western islands of Indonesia? Wallace deduced they had once been part of Asia, and so shared the tropical fauna — monkeys, tigers, and so forth — of Asia.

Wallace was right. The distance between Bali and Lombok is short, but the ocean separating them was later discovered to be very deep. Bali lies just on the edge of the continental shelf, while Lombok lies just off it (see Figure 2.4). Bali was once connected to the other western islands of Indonesia, but never to Lombok. For animals living on one island, it wasn't simply a matter of flying or swimming the twenty miles to the next island. For millions of years the separation was much greater, and so animals adapted to the conditions peculiar to each island. The islands are close together today, but they are, geologically speaking, "new neighbors."

Wallace had linked the question of the origin of species to how species were distributed. He had defined a dividing line between the fauna of Asia and Australia. His discovery would forever be known as the "Wallace Line" (see Figure 2.4), and Wallace himself would be known as the founder of biogeography.

Survival of the Fittest

The question for Wallace was now not if species evolved, but how? Baking in a malarial fever on the volcanic island of Ternate in early 1858, the answers came to him.

Alternating between hot and cold fits, Wallace had to rest. He had nothing to do, he recalled, but "to think over subjects then particularly interesting to me." Wrapped in a blanket on an 88°F day, he thought of Malthus' essay on population, which he had read some years earlier. It occurred to him that maladies similar to disease, accidents, and famine that check the growth of human populations check the populations of animals, too. He thought about breeding, how animals bred much more rapidly than humans, and that if left unchecked, animals would overcrowd the world very quickly. But all of his experience revealed that animal populations were limited. "The life of wild animals," Wallace concluded, "*is a struggle for existence. The full exertion of all their faculties and all their energies is required to preserve their own existence and provide for that of their infant offspring.*" Wallace knew from a decade in the jungle that finding food and escaping danger ruled animals' lives — and the weakest were weeded out.

Wallace, the great collector, was intimately familiar with the variety of individuals of a species. He continued, "Perhaps all the variations . . . must have some definite effect, however slight, in the habits of or capacities of the individuals . . . a variety having slightly increased powers . . . must inevitably acquire a superiority in numbers."

Bingo. He had figured it out — either that or he was out of his mind. Wallace had to wait for his fever to fade before he could make any notes. Then, he wrote the paper out in full in just a few nights.

He entitled the paper, "On the Tendency of Varieties to Depart Indefinitely From the Original Type." Later he would refer to the main idea expressed in this paper as "survival of the fittest" (a phrase borrowed from the social scientist Herbert Spencer). Wallace's paper was just a sketch, conceived in a dilapidated house in an earthquake-ravaged island during bouts of fever, ten thousand miles from the center of science in England. Wallace did not send it directly to a journal, he wanted others to look at it first.

He sent it to, whom else? Darwin.

This time, Wallace would not go unnoticed.

Priority and Posterity

Darwin received Wallace's paper sometime in June 1858. He was shocked when he read it. He should not have been, had he been paying closer attention to the previous dispatches of his faraway correspondent. Nevertheless, Darwin, sixteen years past the first version of his long essay on species formation, quickly wrote to his friend Charles Lyell fearing that "all of my originality, whatever it may amount to, will be smashed."

What happened thereafter is still a subject of debate among scholars. The facts are that Wallace had asked Darwin to forward the manuscript to the geologist Sir Charles Lyell, which Darwin did. Lyell, and J. D. Hooker, the eminent botanist, were intimates of Darwin to whom he had divulged his theory of natural selection and much of the argument supporting it. Lyell and Hooker took the initiative to arrange for Wallace's paper, and a brief sketch from Darwin on his theory, to be read together at an upcoming meeting of the Linnaean Society and for both to be published together.

Was Wallace robbed of his individual right to glory? Was the arrangement of joint publication fair? (Wallace was not informed of it until after the fact.) On the other hand, it was Darwin who coined the term "natural selection" and he had shared his 1842 sketch, at least privately, with other scientists.

It is true that Darwin's name and works are far better known than Wallace's today, but consider Wallace's perspective on the matter. He was always, for the rest of his life, deferential to Darwin. Wallace always referred to the "Darwinian theory" and dedicated his major book about his travels, *The Malay Archipelago* (1869), to Darwin: "[To] Charles Darwin, author of the *Origin of Species*, not only as a token of personal esteem and friendship but also to express my deep admiration for his genius and his works." Wallace was one of the pallbearers at Darwin's funeral, and in Wallace's autobiography, *My Life* (1905), he devoted a full chapter to his friendship with Darwin. Not a word of regret, envy, or resentment.

Perhaps for Wallace it was simply a matter of being accepted. He was, up until 1858, an outsider to the circle of eminent scientists who led the new revolution in thought. When he heard that Lyell and Hooker had made complimentary remarks about his paper, he wrote to his oldest friend and schoolfellow, "I am a *little* proud." Wallace did not need or seek to be the center of the circle, he just wanted inside.

That, and more, he surely earned.

CHAPTER QUESTIONS

1. Why did Wallace choose to go to the Malay Archipelago?

2. Compare and contrast the animals on Bali, Borneo, and the western islands of the archipelago with those on Lombok, New Guinea, and the eastern islands. What are the differences and why are they important?

3. What observations led Wallace to the idea of a "struggle for existence"? How were Wallace's observations similar or different from those driving Darwin's ideas about natural selection?

For more on this story, go to the *Into The Jungle* companion website at www.aw-bc.com/carroll.

FIGURE 3.1. *The massive Amazon River system.*
The main river and its tributaries span more than fifteen thousand miles. Henry
Walter Bates spent most of his eleven years in the Amazon on the main river, while
Alfred Russel Wallace ventured far up the Rio Negro. Bates found more than
550 species of butterflies at Ega (now Tefé). *Drawn by Leanne Olds.*

THREE

Life Imitates Life

A river always leads to some inhabited place. If we do not meet with agreeable things, we shall at least meet with something new.

—Cacambo in *Candide*, Voltaire (1759)

"It was the best of times, it was the worst of times . . ." So began Charles Dickens' famous novel of 1859, *A Tale of Two Cities*. Returning from the Amazon that same year, Henry Walter Bates could begin his tale with the same words.

The best of times was certainly daily life in a naturalist's paradise:

> I rose generally with the sun, when the grassy streets were wet with dew, and walked down to the river to bathe: five or six hours every morning were spent in collecting in the forest . . . the hot hours of the afternoon . . . and the rainy days were occupied in preparing and ticketing the specimens, making notes, dissecting, and drawing. I frequently had short rambles by water in a small montaria [type of canoe] . . . the neighborhood yielded me . . . an uninterrupted succession of new and different forms in the different classes of the animal kingdom, but especially insects.

The worst of times was perhaps the first year on his own upriver, after separating from his companion Alfred Wallace in March 1850:

> Twelve months elapsed without letters or remittances. Toward the end of this time my clothes had worn to rags: I was barefoot, a great inconvenience in tropical forests, notwithstanding statements to the contrary that have been published by travelers; my servant ran away, and I was robbed of nearly all my money.

Broke, lonely, unsure of his prospects and pressed by letters to return home to the family hosiery business, Bates went fourteen hundred miles downriver to the port town of Pará intending to find a boat home. Yellow fever had ravaged the place, and soon after arriving, Bates too was stricken.

※　　※　　※

But Bates did not go home. He turned around and spent eight more years in the Amazon, eleven years in total. *Eleven years!* Why did he stay, and for so long? How did he bear it?

The answer to the first question is that Bates received a timely dose of fresh funds and a letter from his agent in London saying that the specimens he had sent were being very well received. One new butterfly species, *Callithea batesii*, had been named after him. He changed his mind and renewed his determination to go far upriver as he had originally planned.

The length of his adventure was dictated by the massive scope of the Amazon. With more than one thousand tributaries draining about 2.7 million square miles, the Amazon is the largest of any river system (see Figure 3.1). Bates roamed up and down about two thousand miles of the main river in the course of these years. (The ten largest rivers in the system span fifteen thousand miles.) Travel was almost exclusively by water and was slow — very slow. Powered by only paddles or sails, pounded by storms and rain, and subject to changing winds, or no wind at all, Bates was often a passenger in some small trading vessel plying the river, or a canoe belonging to members of one of the many local tribes. Here's a scene from a typical crossing Bates made so that he could hunt for a monkey on the opposite bank:

> We were about twenty persons in all, and the boat was an old rickety affair . . . In addition to the human freight we took three sheep with us . . . Ten Indian paddlers

carried us quickly across . . . When about half-way, the sheep in moving about, kicked a hole in the bottom of the boat. The passengers took the matter very coolly, although the water spouted up alarmingly, and I thought we should inevitably be swamped. Captain Antonio took his socks off to stop the leak, inviting me . . . to do the same, whilst two Indians baled out the water . . . We thus managed to keep afloat.

By managing only a few miles or so at a stretch, the branches and twigs of the Amazon system would seem endless. Bates wanted to see all of the expansive river's treasures. That passion and determination, and the rewards of dashing into the forest at nearly every bend, seem to have offset the unrelenting heat, malaria, yellow fever, fire ants, biting flies, and intense loneliness that he endured.

Those rewards were many — river dolphins, anteaters, frigate birds, anacondas, hummingbirds, bird-eating spiders, all sorts of monkeys, jaguars, caimans, blue hyacinthine macaws, parrots, eagles, five species of toucans, and butterflies — flocks of butterflies. Bates collected 14,712 animal species in all, of which more than 8,000 were new to science.

Eventually, the grueling work, bad and insufficient food, and overall deterioration of his health convinced Bates to return home to England. The parting was bittersweet:

> On the evening of the third of June [1859], I took a last view of the glorious forest for which I had so much love, and to explore which I had devoted so many years. The saddest hours I ever recollect to have spent were those of the succeeding night, when the mameluco pilot left us free of the shoals and out of sight of land . . . I felt that the last link which connected me with the land of so many pleasing recollections was broken . . . Recollections of English climate, scenery, and modes of life came to me with a vividness I had never before experienced during the eleven years of my absence. Pictures of startling clearness rose up of the gloomy winters, the long grey twilights, murky atmosphere, elongated shadows, chilly springs, and sloppy summers . . . To live again amidst these dull scenes I was quitting a country of perpetual summer . . . It was natural to feel a little dismayed at the prospect of so great a change . . .

Returning in the summer of 1859, Bates' timing was quite fortunate. Within months Darwin's *Origin of Species* would appear and give Bates a concrete framework for thinking about all he had seen and collected.

The Butterflies of Ega

No group of animals made a greater impact on Bates than the butterflies of the Amazon. They were, of course, highly prized back home for their beauty. Because Bates made his living by selling specimens, he paid careful attention to the varieties in each locale he visited.

The variety was overwhelming. In the vicinity of Ega alone, in the Upper Amazon where Bates spent more than four years, he found 550 distinct species of butterflies. This figure towers over the just sixty-six species in all of Britain and the three hundred or so in all of Europe.

The butterflies at Ega and throughout the Amazon presented several puzzles to Bates' expert eye. For example, despite years of experience he could never tell some species of *Leptalidae*[1] from those of *Heliconidae* while they were in flight. Their wing markings were very similar and they flew together in the same parts of the forest. Only upon close inspection after capture could Bates tell from finer details of the wing vein and color patterns which species was which. One species, *Leptalis theonoë*, in its different varieties, resembled several different species of *Ithomia* butterflies.

Bates was very careful in noting where certain varieties were present or absent. He noticed that none of the particular *Leptalis theonoë* types that resembled a particular *Ithomia* species was found in any other district or country. The *Leptalidae* "counterfeits" were only passing themselves off where the real *Ithomia* species existed in abundance. Bates termed this phenomenon "mimetic analogy," or mimicry.

When Bates read *The Origin of Species*, he was one of the few immediate adherents. As he contemplated his butterflies, he realized the significance of mimicry as evidence for the process of natural selection. He struck up a correspondence with Darwin in 1860, just as the firestorm of controversy was erupting over the elder's great book. Bates wrote to Darwin, "I think I have got a glimpse into the laboratory where Nature manufactures her new species."

---

[1] Bates' original term for the group has been changed to *Dismorphia*.

Darwin was positively thrilled. Bates, who had no formal scientific position, was feeling a bit discouraged at the time at not being part of the scientific establishment. For the first three years after returning from the Amazon, he lived back in Leicester with his family.

But Darwin rooted him on. He urged Bates to present his work to the most important scientific societies, to publish in the most influential journals, and to write an account of his journey as a travel narrative, as Darwin had done for his *Beagle* voyage. Bates gobbled up Darwin's advice. Theirs was a warm, symbiotic friendship.

One of the Most Remarkable and Admirable Papers

Bates set to work on both the formal scientific description of his Amazon collections and a book on his travels. Both were massive tasks. Bates would later remark that he would rather spend another eleven years in the jungle than go through the ordeal of writing another book.

But the same discipline that brought him success in the jungle brought him success as a scientist and author. His most important paper, with the misleadingly dull title, "Contribution to an Insect Fauna of the Amazon Valley, Lepidoptera: Heliconidae," laid out the evidence and a mechanistic explanation for the phenomenon of mimicry.

Bates noted that several species of *Dioptis*, a genus of moths, also mimic species or local varieties of *Ithomia* butterflies. He explained that a series of mimetic relationships also occurred in the Old World, between Asiatic and African *Danaidae* butterflies and species of other families of butterflies and moths. Most importantly, he underscored that no instance was known in these families of a tropical species of one hemisphere counterfeiting a form belonging to the other hemisphere. In other words, these were not accidental resemblances of butterflies with different ranges, the mimicry occurred among species found in the same area (Figure 3.2).

Furthermore, Bates knew firsthand that mimicry occurs among other insects. Along the banks of the Amazon he found parasitic bees and flies that

FIGURE 3.2 *Mimicry in butterflies.*
This is an original plate from Bates' 1862 paper reporting the discovery of mimicry.
The butterfly at the center (5) is *Leptalis nehemia,* the typical butterfly of the family.
The other *Leptalis* butterflies (1–8) deviate greatly from this pattern, as they are
mimics of other species. Each pair (3/3a, 4/4a, 6/6a, 7/7a, 8/8a) illustrates mimicry
between *Leptalis* and species of other families. Specimens 3a, 4a, and 6a are members
of the genus *Ithomia* that mimic varieties of *Leptalis theonoe* found in the area of Sao
Paulo. Specimens 7a and 8a are members of the *Mechanitis* and *Methona* genera that
mimic *Leptalis amphione* and *Leptalis orise.*

FIGURE 3.3 *Caterpillar mimic of snake head.*
First discovered by Bates, a number of species mimic the appearance of snake
heads. This is the Spicebush Swallowtail caterpillar (*Papilio trollus*).
Photo by Mary Jo Fackler.

do not build nests but mimic the forms of nest-building bees to live "all
expenses paid" in their nests. He found a cricket that was a good imitation of a
tiger beetle and always found on trees frequented by the beetle. The most striking example of imitation was a very large caterpillar he found that, when
stretched in the foliage of a tree, startled him by its resemblance to a small
snake. The caterpillar had black spots on segments of its head which when
expanded resembled the head of a viper (Figure 3.3). When Bates carried the
specimen into the local village it frightened everyone who saw it.

Bates saw mimicry in a Darwinian light and proposed that the specific mimetic forms of insects were adaptations. He witnessed, in the Amazonian jungle, that every species maintained its existence by virtue of some traits that enabled it to withstand "the battle of life." He knew hundreds of examples of how animals conceal themselves from their enemies. Clearly, one species disguising itself as another was one of these self-preservation strategies, or in Bates' words, "The adaptive resemblance of an otherwise defenseless species to one whose flourishing race shows that it enjoys particular advantages."

The advantages of mimicking a poisonous snake were obvious, but what advantages did the *Heliconidae* butterflies possess that made them so abundant, and the objects of imitation? It was not obvious what might help them escape the many insect eaters of the forest. Bates, though, had a good and ultimately correct idea. He knew too well that some butterflies secreted foul-smelling fluids and gases when he handled them. He noticed that when he laid such specimens out to dry, the various jungle vermin were less likely to carry them off. Bates also never saw the flocks of slow-flying *Heliconidae* pursued by birds or dragonflies, for which the butterflies would be easy prey. Nor when resting were the *Heliconidae* attacked by lizards or predacious flies, which pounced on other butterflies. Bates surmised that the *Heliconidae* must be unpalatable, and that other palatable species disguise themselves by mimicking the wing patterns of the *Heliconidae* and are thereby protected from predators.

Bates saw the origin of mimicry in general terms — as the same process that involves the origin of all species and adaptations. The case of *Leptalis theonoë* and *Ithomia* was most telling. The form of the *Leptalis theonoë* species in each region depended on the form and colors of the *Ithomia* butterflies in each region, which varied from place to place. Bates asked in his paper: How are local races formed out of the natural variations of a species?

> The explanation of this seems to be quite clear on the theory of natural selection, as recently expounded by Mr. Darwin in the 'Origin of Species' . . . If a mimetic species varies, some of its varieties must be more and some less faithful imitations of the object mimicked. According, therefore, to the closeness of its persecution by enemies, who seek the imitator, but avoid the imitated, will be its tendency to become an exact counterfeit, the less perfect degrees of resemblance being, generation after generation, eliminated, and

only the others left to propagate their kind . . . To exist at all in a given locality, our *Leptalis theonoë* must wear a certain dress and those of its varieties that do not come up to the mark are rigidly sacrificed . . . I believe the case offers a most beautiful proof of the theory of natural selection.

So did Darwin. He called it "one of the most remarkable and admirable papers I ever read in my life," and he assured Bates, "It will have lasting value." For Darwin and other proponents of natural selection, Bates' work was a powerful display of the process. Darwin's evidence in *The Origin of Species* had relied heavily on the analogy of natural selection to the domestication of animals. With Bates' observations, Darwin had rich, independent evidence of natural selection in nature.

Argument and Evidence

Mimicry became a focal point of the debate between proponents and opponents of the theory of natural selection. The debate was, for several decades after the publication of Bates' paper, mostly a matter of different interpretations of the same observations. Of course, the better approach was to gather more evidence that would weigh for or against natural selection as a cause. Such evidence has unfolded as biologists studied mimicry in greater depth.

One of Bates' key inferences was that the species being mimicked is unpalatable to predators and that the palatable species gains protection by imitating the unpalatable species' appearance. This would imply that predators learn or inherently know to avoid the unpalatable form.

There were many subsequent investigations of this suggestion, but the most notable controlled studies were conducted by Jane van Zandt Brower beginning in the late 1950s. Using wild caught birds, Brower showed that many butterfly species thought to be unpalatable were rejected and avoided by birds. Furthermore, birds showed a tendency to learn rapidly from experience to recognize unpalatable butterflies, and to reject those resembling them.

A second key aspect of mimicry is the expectation that protection from predators should break down where the unpalatable form is not present. This prediction has recently been tested in a fascinating example of mimicry among snakes.

FIGURE 3.4 *Mimicry in snakes.*
Arizona Mountain kingsnake (*top*) and Arizona coral snake (*bottom*).
Photos by Gary Nafis.

The harmless and very beautiful scarlet kingsnake and Sonoran mountain kingsnake superficially resemble poisonous coral snakes in that each species possess red, yellow, and black ring patterns (Figure 3.4). The sequence of color bands differ between the harmless and poisonous snakes according to the helpful rhyme memorized by many snake enthusiasts: "Red touch yellow, kill a fellow. Red touch black, friend of Jack."

David and Karin Pfennig and William Harcombe of the University of North Carolina (Chapel Hill) identified dozens of sites in North Carolina, South Carolina, and Arizona where both coral snakes and king snakes occur together, and where the coral snake does not occur. At each site, they left ten sets of three snake replicas made of cylinders of plasticene (a soft, non-toxic clay): one with a tricolor ringed pattern, one with a striped pattern of identical colors, and one with a plain brown pattern. After weeks in the field, the plasticene replicas were collected and scored for predator bite and scratch marks by an individual who did not know whether the replicas had been placed in areas where the snakes overlap or where the poisonous species was absent.

It was found that at the Carolina sites, the proportion of kingsnake replicas attacked was high (68 percent) in areas where coral snakes do not occur, but low (8 percent) in areas where coral snakes do occur. Similar results were found at the Arizona sites, confirming that predators avoid coral snake mimics in areas where coral snakes live.

As Darwin predicted, Bates' work has had lasting value and biologists to this day refer to the imitation of unpalatable or poisonous forms by palatable, harmless species as Batesian mimicry.

＊　＊　＊

Bates never returned to the Amazon, but he did manage to finish the book of his travels, *The Naturalist on the River Amazons* (1863). When it was published, Bates sent a copy to Darwin and anxiously awaited the verdict from the famous author of *The Voyage of the Beagle*. Darwin replied, "My criticisms may be condensed into a single sentence, namely, that it is the best work of Natural History Travels ever published in England."

It is still a great read today—full of tales of adventure and great descriptions of the animal and human residents of the Amazon. When describing butterfly wings, which became to Bates what the Galapagos finches represented to Darwin, Bates achieved the level of poetry. He wrote, "It may be said, therefore, that on these expanded membranes nature writes, as on a tablet, the story of the modifications of species."

CHAPTER QUESTIONS

1. What was the "glimpse" that Bates had "of how nature manufactures her species?"

2. What evidence did Bates assemble to argue that mimicry was due to natural selection and not mere coincidence?

3. Why was Darwin so delighted by Bates' discovery of mimicry?

For more on this story, go to the *Into The Jungle* companion website at www.aw-bc.com/carroll.

PART II

On the Trail of
Ancient Humans

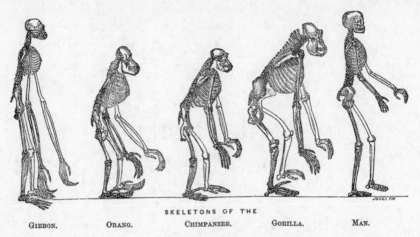

SKELETONS OF THE

GIBBON.   ORANG.   CHIMPANZEE.   GORILLA.   MAN.

FIGURE 4.1 *The evolution of apes and humans.*
The famous frontispiece from T. H. Huxley's *Evidence as to Man's Place in Nature* (1863), which stressed the kinship of humans and apes.

Java Man

"No great discovery was ever made without a bold guess."

—Isaac Newton

His colleagues thought that he had lost his mind.

What else could explain why this promising young physician and anatomist, sure to rise to a prestigious professorship at the leading medical school in Holland, would throw it all away for a posting with the Dutch Army ten thousand miles away in the East Indies? What's more, how could he in good conscience take his beautiful young wife and new baby away from their home to such a faraway, foreign, and dangerous place?

What did twenty-nine-year-old Doctor Marie Eugène François Thomas Dubois, born the very year that Wallace's and Darwin's first papers on natural selection were announced to the world, hope to find in the East Indies?

Dubois sought the most important discovery he imagined that anyone could make in these early days of evolutionary theory — the "missing link" between apes and Man. Definitive proof of the link between humans and the rest of the animal world would be the ultimate crown on the theory of evolution, and the name Eugene Dubois would be famous and respected all over the world.

* * *

The journey was, in Dubois' mind at least, not a flight of fancy, but a matter of destiny. Dubois believed that everything he had learned, from his boyhood exploration of the plants and fossils in his native Limburg, to his early schooling and later medical training, to his current anatomical research on the human larynx, had prepared him for this great quest.

As a child, Dubois was absorbed in the study of nature. He made frequent trips into the countryside to collect medicinal herbs for his father's pharmacy. From his bedroom he could see St. Peter's mountain near Maastricht, a rich fossil chalk bed famous for the discovery in 1780 of the first Mosasaur (from the Latin *Mosa*, meaning the "Meuse river" in the Netherlands, and the Greek *sauros*, meaning *lizard*), a late Cretaceous marine reptile. Dubois went on many fossil-collecting expeditions into the vast chalk formations.

From his early school days on, the question of human origins swirled about him. When he was just ten years old, Dubois heard about a series of lectures by zoologist Carl Vogt that stirred up controversy throughout Holland. A staunch supporter of Darwin's new theory, Vogt embraced the view of humans as being members of, and not above, the animal kingdom. In high school, Dubois' science master introduced him to the great works by Darwin, Thomas Huxley, and Ernst Haeckel that developed and promoted these ideas.

Darwin and Wallace had solved the "mystery of mysteries." Dubois became focused on what Huxley described in *Evidence as to Man's Place in Nature* (1863) (see Figure 4.1) as, "The question of questions for mankind — the problem which underlies all others, and is more deeply interesting than any other — is the ascertainment of the place which Man occupies in nature and of his relations to the universe of things."

Huxley's book was the first detailed biological examination of humans in the light of emerging comparative studies of apes and of the first fossil humans, which Huxley had pursued intensively. While Darwin had stated in *The Origin of Species* that "light will be thrown on the origin of man and his history,"

FIGURE 4.2 *Eugene Dubois, age twenty-five.*
Photo courtesy and copyright of the National Museum of Natural History,
Leiden, the Netherlands.

he deliberately avoided any further discussion of human origins, figuring
(correctly) that his great theory had so many objections to overcome without
bringing the delicate issue of human evolution into the fray.

But Huxley picked up where Darwin left off and confronted the central issues
head-on. Huxley pleaded for a dispassionate, objective zoological approach.
He asked his readers to approach human biology as someone from another
planet would:

> Let us . . . disconnect our thinking selves from the mask of humanity; let us
> imagine ourselves scientific Saturnians, if you will, fairly acquainted with such
> animals as now inhabit the Earth, and employed in discussing the relations they
> bear to a new and singular 'erect and featherless biped,' which some enterprising
> traveler, overcoming the difficulties of space and gravitation, has brought from
> that distant planet for our inspection, well preserved, may be, in a cask of rum.

He then asked:

> Is man so different from any of these Apes that he must form an order by
> himself? Or does he differ less from them than they differ from one another,
> and hence must take his place in the same order with them?

Huxley urged his readers:

> Being happily free from all real, or imaginary, personal interest in the results of
> the inquiry thus set afoot, we should proceed to weigh the arguments on one
> side and on the other, with as much judicial calmness as if the question related
> to a new Opossum.

Dubois saw that such "judicial calmness" was in short supply.

Huxley's analysis of human bodies, brains, and eggs built the zoological
case for the connection between man and other animals, but he also brought
then newly available paleontological evidence into the argument. The first
Neanderthal remains, and a second fragmentary skull found in Belgium along
with remains of mammoths and the woolly rhinoceros, were, to Huxley, definitive evidence of human antiquity.

Others inside and outside of the science community were skeptical, if not
outright hostile to the idea of ancient humans. The leading German pathologist Rudolf Virchow concluded that the Neanderthal's unique skeletal features
were deformities caused by disease, and the remains were not evidence of a distinct human race or species.

Dubois' interest in the "question of questions" was further stoked by the ideas
of German embryologist Ernst Haeckel. In his *History of Creation* (1868),
Haeckel outlined a speculative history of human origins, beginning with a simple single-celled ancestor. Building upon Huxley, Haeckel underscored what he
thought were the two most important adaptations that made humans distinct —
walking upright and articulate speech. These capabilities were made possible,
Haeckel asserted, by two major morphological changes — "the two pairs of
limbs and the differentiation of the larynx." Haeckel proposed that walking
upright long preceded the acquisition of speech and that there was a stage in the
evolution of humans, which he called, "Speechless Man (*Alalus*), or Ape-Man
(*Pithecanthropus*), whose body was indeed formed exactly like that of a Man in
all essential characteristics, but did not as yet possess articulate speech."

Neither Haeckel nor Huxley viewed Neanderthal as an intermediate between "Men and Apes," and Huxley closed his book asking:

> Where, then, must we look for primaevel Man? . . . In older strata do the fossilized bones of an Ape more anthropoid [human-like], or a Man more pithecoid [ape-like], than any yet known await the researches of some unborn paleontologist?

> Time will show.

That paleontologist was not unborn at the time Huxley wrote those words, but he was just five years old. And later, each time he read those words, over and over again, his determination to be that paleontologist grew.

Amsterdam

Dubois' father wanted his son to follow in the family business and become a pharmacist. But Dubois, a fiercely independent young man, was determined to continue his studies of nature. That meant going to medical school, where the first year of studies was focused entirely on the natural sciences. In 1877, at age nineteen, Dubois enrolled at the University of Amsterdam.

The faculty was outstanding and included luminaries such as the physicist Van der Waals (Nobel Prize in Physics, 1910), the chemist Van't Hoff (winner of the first Nobel Prize in Chemistry), and botanist Hugo de Vries (one of the rediscoverers of Mendel's work on heredity). De Vries and Dubois often discussed the raging debate over human origins.

Dubois soon realized that he had little interest in becoming a practicing doctor, but he did not take Darwin's way out. Dubois worked hard, was disciplined, very focused, and did very well at every subject he touched. His talent was recognized, and in 1881 Dubois was offered an assistantship in anatomy by Dr. Max Fürbringer. It was a stroke of luck, as Fürbringer was himself trained by Haeckel. Dubois excelled, and Fürbringer helped Dubois gain promotions, from assistant to prosector in charge of the anatomy course, to lecturer, all in rapid succession. It was a meteoric rise. Dubois was just one rank below full professor at the tender age of twenty-eight.

Dubois decided to start up his own independent research program, on the comparative anatomy of the larynx, the structure responsible for the unique human capability of articulate speech. He published one paper, but a series of experiences soon turned him away from his budding academic career and onto a course for the East Indies.

The first blow to Dubois' academic ambitions was his discovery that he hated teaching. He was so anxious and wound up before lectures that he talked to no one, and wanted none of his colleagues to attend them. The second blow was a falling-out with Fürbringer. Dubois was very ambitious and eager to gain credit and recognition for his work. When he gave a draft of a manuscript for his first article on the larynx to Fürbringer, the latter commented that he had previously made some of the points Dubois was arguing. Dubois worried that his work would not be seen as his own. He revised the article, but never stopped stewing over Fürbringer's comments and grew increasingly suspicious of Fürbringer's motives.

The third catalyst to Dubois' break from his career path was new fossil finds, which reignited his interest in human paleontology. In 1886, near Spy, Belgium, more Neanderthal remains were found. There was no doubt these were old bones, and they demolished the argument that the original find in Germany was that of a diseased individual. No, the German and Belgian Neanderthal fossils were something different from modern humans, but still a long way from apes or an ape-like ancestor.

Dubois worried that the years were flying by and others were getting close to finding the missing link. If he was to seize the prize he had to take action.

Where to?

Dubois made up his mind to chuck it all — the certainty of a professorship, the teaching, the anatomy lab, Fürbringer — to look for the missing link. The trouble was — where to look?

Certainly not in Europe, where the Neanderthals had been found. Darwin had suggested in *The Descent of Man* that because humans lost their fur covering, they must have originated in the Tropics and not colder zones. That excluded North America, but left Africa, Asia, part of Australia, and South America as possibilities. But, because apes were found only in tropical parts of the Old World, it followed that human ancestors originated in those same regions — that left Africa or Asia.

Dubois knew that Darwin favored Africa because of humans' affinity to gorillas and chimpanzees. But Asia held the gibbon and orangutan, and Haeckel had argued that gibbons were more closely related to humans. Furthermore, a fossil ape known as the Siwalik chimpanzee had recently been discovered in the Siwalik Hills of British India.

The age and location of this find suggested that deposits of similar age might be fruitful. Studies by a Dutch paleontologist indicated that such deposits might be found in Borneo, Sumatra, or Java. Dubois was well aware of Wallace's "line" and Wallace's work on the geographic distribution of animals (see Chapter 2). He knew that the animals of the western part of the Malay Archipelago were shared with mainland Asia, so whatever might be found in India could occur in those islands.

Furthermore, all human fossils that had been found to date were found in caves, and Sumatra was littered with caves. And there was one very practical factor pushing Dubois to Sumatra — it was part of the Dutch East Indies. He would find fellow countrymen and some familiar customs — perhaps he might even get some government support for an expedition.

Dubois made a pitch to the secretary-general of the Colonial Office. He laid out his logic for the expedition and the glory the discovery of the missing link would bring to Dutch science. But the secretary-general explained that there was no money to support such a speculative venture.

Dubois had to support himself and his family, so what could he do? He did have marketable skills that were needed in the Dutch Colonies — he was a physician. The Dutch Army needed him, so he enlisted, which was *an eight-year commitment*.

When he told his wife, Anna, she was surprisingly supportive. But his parents and in-laws were not. His disapproving father saw him throwing away a distinguished career. But there was no stopping Dubois now. He and his wife and daughter (Marie) would sail to Sumatra (see Chapter 2, Figure 2.2 for map).

Sumatra

The voyage from Amsterdam to Padang, Sumatra, took forty-three days, even with the shortcut through the Suez Canal. Dubois and his family arrived on December 11, 1887 and tried to settle in among the exotic sights and scents of Sumatran life. It was the rainy season, and Dubois and Anna quickly learned what that meant — buckets of rain every day and mud everywhere. Anna, pregnant with their second child, tended to the setting up of the household, while Dubois reported to the army hospital.

Conditions in the hospital were unlike anything Dubois had seen in Holland, or imagined. He was overwhelmed with patients suffering all sorts of fevers from cholera, malaria, typhus, tuberculosis, and other unidentified maladies. The workload was so heavy that he was not sure when, or if, he may get into the field to prospect for bones.

Dubois made do for a while. He told his fellow officers about his reasons for being in Sumatra — the missing link — and put together a lecture explaining his logic. He used that lecture as an outline for an article that he would write for the *Journal of the Natural History of the Netherlands Indies*. The article served to stake his claim to his search, and as a not-so-indirect warning to the government authorities — support this scientific work or see the glory reaped by another nation.

Dubois' forays into the countryside around the base were not promising, so he requested and received a transfer to another hospital in a more remote region, with more caves to look into and fewer patients to look after. Anna, at eight months pregnant, had to start all over in putting together a new home, but the move to the cooler highlands was a relief from the swelter of Padang. She gave birth to their first son at home in Pajakambo attended to by her husband.

The lighter load at work gave Dubois more time to search, and although the fossil finds were spotty, he had some luck. He found a cave called Lida Adjer that had a good many bones of rhinos, pigs, deer, porcupines, and other

Pleistocene animals. In the meantime, his article drew the attention of the governor, who promised him some laborers for his explorations. Dubois also wrote the governor about his new finds. Even a couple of colleagues back home were supporting Dubois' work and encouraging government support for it.

In March 1889, the government authorized the assignment of two engineers and fifty laborers to help with the search and excavations of the very large number of Sumatran caves. Now, at last, Dubois thought, he can work properly. Surely, finding the missing link was just a matter of time.

But many of the caves were empty, or inhabited by animals that were far from fossilized. One day, Dubois was frustrated at the workers' reluctance to enter one cave, so he took the lead and crawled into its narrow passageway. Penetrating further, he was overwhelmed by the smell of cat urine and rotting meat — the cave was a tiger's lair. Dubois tried to back out quickly, but became stuck and had to plead with his workmen to pull him out.

Dubois shook off the episode, but he couldn't shake off other dangers in this land. He was felled by a bout of malaria, the first of many that would interrupt his work, if they didn't kill him outright. Malaria soon felled many workmen. One of the two engineers then died of fever, and half of his workmen were too sick to excavate the caves. Others quit and ran off. Months passed without success and Dubois, now two years in Sumatra, wrote to the director of the National Museum of Natural History in Leiden:

> Everything here has gone against me, and even with the utmost effort on my part, I have not achieved a hundredth part of what I had visualized . . . I have found a few very useful caves, but still never the best one could wish for. What's more, it was necessary to live out in the forest for weeks on end, usually under an overhanging rock or in an improvised hut, and it turns out that in the long run I can't stand up to that, however well I was able to bear the fatigue at first. Having now come back, with my third bout of fever, which nearly finished me . . . I have had to give it up for good . . . .

Dubois began to rethink his plans. He had heard from a geologist that the fossils on the island of Java could be older than those he was finding on Sumatra. Moreover, a fossilized human skull had been found there the previous year in a rock shelter, a hopeful sign that the caves and shelters on Java might be more productive. He put in for and received a transfer to Java.

Java Man

After more than two years on Sumatra, the family, now four in all, packed up and left for Java. They found a good house in the town of Toeloeng Agoeng and settled in. Far from any army base, Dubois was free to pursue science full-time. He was assigned a new crew, led by two corporals who were far more competent than his engineers on Sumatra.

Dubois started his excavations at Wadjak in June 1890, where the human fossil skull had been found two years earlier. He was quickly rewarded. His crew found all sorts of extinct mammals including rhinos, pigs, monkeys, antelope, and even another partial human skull.

Dubois then made a decision that would prove critical. He expanded the search beyond the caves and rock shelters in the hills, to along the riverbanks. In the dry season, when the waters were low, the sediments of the riverbanks were exposed (Figure 4.3). In the hills and along the banks of the Solo River, Dubois' team uncovered unusually rich deposits that included more rhinos and pigs, hippopotami, two different types of elephant, big cats, hyenas, crocodiles, and turtles.

Then, on November 24, 1890, a fragment of a human jaw with two teeth in it was found. Its decrepit state made further identification difficult, but it was an encouraging sign for the next year's excavating season.

The bones were piling up on the veranda at home (Figure 4.4). Dubois wanted to write up descriptions of the fossils and the places where they were found, but the magnitude of the task was overwhelming, and still growing. And he did not yet have his prize, if it was to be found in Java, or anywhere else for that matter.

During Dubois' second season on Java, digging began at Trinil, on the Solo River. In September 1891, workers unearthed the third molar tooth of some primate. Dubois figured it for a chimpanzee like that of the Siwalik Hill type.

The Trinil site was rich and produced many more mammal fossils. The next month, in October, the engineers unearthed a bone they took for part of a turtle shell. Concave and a deep brown color, the fossil was sent to Dubois at his home.

FIGURE 4.3 *Excavation at Trinil on Solo River.*
Photo circa 1900, courtesy and copyright of the National Museum of Natural History,
Leiden, the Netherlands.

FIGURE 4.4 *The fossils piling up on Dubois' veranda.*
Photo courtesy and copyright of the National Museum of Natural History,
Leiden, the Netherlands.

FIGURE 4.5 *The skullcap found at Trinil.*
The skullcap measured about 18.5 cm long by 13 cm wide. *Photo courtesy and copyright of the National Museum of Natural History, Leiden, the Netherlands.*

There was no doubt this was not a piece of shell, but part of a skull, a primate skull. It had a brow-ridge like that of a chimpanzee (Figure 4.5), but the skull seemed to have encased a larger brain than that of a chimp. It appeared to be from some kind of ape. Dubois would need some other skulls to make some detailed comparisons, and so he requested a chimpanzee skull from Europe.

Dubois was anxious to find more fossils, but the dry season drew to a close. He spent the winter cleaning the skullcap and securing other skulls so he could make some detailed comparisons between them and his new find.

By May 1892, excavation work resumed at Trinil. First, the silt from the long rainy season had to be removed before the workers could start digging through the plot that Dubois and his engineers had staked out. Dubois spent more time at the site. But, by the end of July, after a two-week stint, he was worn out. He

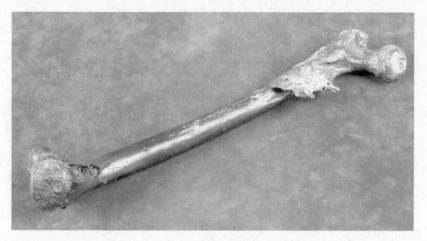

FIGURE 4.6 *The thigh bone found at Trinil.*
This fossil was very similar to that of modern humans, leading Dubois to conclude
that *Pithecanthropus* was well over five feet tall. *Photo courtesy and copyright of the
National Museum of Natural History, Leiden, the Netherlands.*

wrote in his diary: "I discover that there is no more unsuitable place available
in Java, because of health and malaria, for the study of fossils than this hell . . ."

Dubois developed another bout of fever.

The next month, the engineers again made a remarkable find, this time an
almost complete left thigh bone. When the piece reached Dubois he was
delighted. He could tell that this animal was in no way equipped to climb
trees. It was very human-like (Figure 4.6).

Now he had a molar, a skullcap, and a thigh bone. It made perfect sense to
Dubois that these fossils, found relatively near one another, though at separate
times, came from one individual. The thigh bone was crucial to Dubois. Its
features indicated that it came from an upright walking ape, and therefore was
a new species. He named his find *Anthropopithecus erectus* Eug. Dubois —
the upright walking chimpanzee.

But he soon discovered he had made a mistake — a wonderful mistake.
When he first estimated the volume of the braincase from the skullcap, he
obtained a figure of 700 cc, larger than a chimpanzee (410 cc), but much
smaller than a human (~1,250 cc). But he had measured the skullcap and

figured the braincase capacity incorrectly. A recalculation put the skullcap closer to 1,000 cc, much larger than any ape, and much closer to, but not fully that of a modern human. His fossil was not an ape, not a human, but an upright-walking *intermediate* between apes and humans.

He had done it.

Five years after arriving in the East Indies, after leaving his job, parents, and Holland, and after scouring countless caves, dodging tigers, and battling bout after bout of malaria, he had found the missing link.

He renamed his find *Pithecanthropus erectus* — erect ape-man. It was time to tell the world.

The World Reacts

If there were a Hollywood version of this story (and maybe there ought to be!), the film would end here and we could all walk out of the theatre smiling, having seen that Dubois' bold gamble, damaged health, hard work, and his family's many sacrifices were rewarded with great luck and the coveted prize. And we could be certain that fame and the acclaim of the scientific world was sure to follow.

But that is not what happened. To claim such a prize as the missing link, Dubois and his fossils would have to withstand a storm of critical scrutiny; some of it good, honest scientific analysis as it should be, and some not.

For most of 1893, Dubois worked on putting together a description of *Pithecanthropus*. He first thought he could write a series of articles on his work in Java, with the ape-man as part of it. But that would require dealing with tens of thousands of fossils that had been collected. He soon decided that his prize specimen must get his full attention.

Dubois' thirty-nine-page description included photographs of the thigh bone and skullcap, and comparative illustrations of other ape skulls. Dubois emphasized the close proximity of the places where the molar, skull, and thigh bone were found and expressed his strong conclusion that the remains were all from the same individual. He examined details of the skull and pointed out both its several human-like and ape-like features, and its large capacity.

Dubois argued on the basis of the thigh bone that *Pithecanthropus* walked in the way humans do, and was about the same height and body size. Altogether, his ape-man was just that — something in-between ape and man. He stated: "*Pithecanthropus erectus* is the transitional form which, according to the theory of evolution, must have existed between Man and the anthropoids; he is Man's ancestor."

Dubois had the work printed in Batavia, the capital of the Dutch East Indies, and it reached Europe by the end of 1894.

Dubois eagerly awaited reactions from Europe. He did not have to wait very long, but the reactions were not what he expected or hoped for. Criticism poured in from many directions. A German anatomist declared the skullcap was undoubtedly that of an ape, and the femur was human. He credited Dubois only with finding a fossil gibbon, and more evidence of the antiquity of humans. Rudolf Virchow, the outspoken skeptic on Neanderthals, also concluded that the skullcap was that of a gibbon. He refused to accept *Pithecanthropus* as a missing link.

Other commentators took a different angle. Some British scientists saw the skull as human. The paleontologist who described the Siwalik chimpanzee wrote a critique in the prominent journal *Nature* that suggested that the skull was that of a human with microcephalic disease (a condition that arrests brain and skull growth). Some suggested that the bones could have come from a member of a primitive human race, and were not a transitional form.

There were some exceptions to the skeptics. American paleontologist O. C. Marsh thought that Dubois had proven his case. And Haeckel was, not surprisingly, supportive.

But the majority of opinions were negative. They stung. Dubois fumed. Here he was halfway around the world, living and working in primitive conditions, making actual discoveries, while these European academics were sitting in the comfort of their lofty offices. They had some gall writing papers and giving lectures on fossils they had never seen, and doubting the analysis of the one man who had found and studied those fossils! Dubois concluded that the ferocious criticism must be caused by jealousy. He had found the missing link, and now they wanted to deny him his rightful credit!

He had to go back to Europe and convince the doubters.

Home

The Dubois family had nearly eight years' worth of possessions to sort through to decide what was to be shipped home, what they would carry with them, and what would stay behind in Java. Dubois had 414 crates containing more than twenty thousand fossils, but there was only one possession on his mind, his *Pithecanthropus.* He had a special wooden suitcase made to hold the two wooden boxes that cradled the precious fossils. He would carry this himself to Batavia, on the ship to Marseilles, and on the long train ride to Amsterdam.

It was a six-week journey from Java to Europe and Dubois used the time to prepare, both strategically and psychologically, for the battle ahead. But nature decided he must endure one more challenge. Out on the Indian Ocean, the ship was engulfed in a terrific storm. The captain ordered the passengers into the lifeboats, just in case they had to abandon ship.

Just then, Dubois remembered that he had left *Pithecanthropus* in his cabin. If he were to lose that, he would have nothing to show for all of his efforts, and nothing with which to defend himself against the backlash brewing in Europe. He went back to get it and told Anna that if the boats were lowered, she was to look after the children. He would hold on to his newest child — *Pithecanthropus.*

Fortunately, the storm passed and no such separation was necessary. By early August, the family reached Holland.

For Dubois, it was not a triumphant reunion. His father passed away while he was in Java, so he would not have the satisfaction of proving that his long journey to Indonesia was worth all the sacrifice. And his mother was not impressed by her son's box of bones. Seeing *Pithecanthropus,* she asked, "But, boy, what use is it?" Some of Dubois' critics were asking the same question.

* * *

Dubois hurled himself into a campaign to persuade all of Europe of the importance of his discovery and the correctness of his interpretation. He and *Pithecanthropus* commenced a tour of scientific conferences and the most important institutions in Germany, France, Belgium, and Great Britain.

His first opportunity arrived in just a few weeks, at the International Congress of Zoology being held in Leiden. Many key figures were present, including arch-critic Virchow, who presided over the meeting. Dubois knew he must be in top form to stand up to Virchow's unwavering skepticism, and outright ridicule. Earlier that year, Virchow had described Dubois' interpretation of the Java fossils as a "fantasy . . . beyond all experience."

Dubois wisely avoided any personal attacks and focused instead on the legitimate scientific questions that had been raised about the fossils. He acknowledged that his thirty-nine-page description was inadequate on some important matters, and he attempted to fill in those gaps. He described more fully the geological formation in which the fossils were found, and the other animal bones located in the same layers (Figure 4.7). He emphasized the human-like characteristics of the thigh bone and the ape-like features and intermediate braincase capacity of the skullcap.

Virchow was unmoved, but others acknowledged that Dubois was clearing up some of the misconceptions and unanswered questions from his original description.

Most important to his campaign, Dubois allowed many potentially influential scientists to see and inspect the bones for themselves. That warmed a few to his side. In Paris, he developed a great ally. He traveled to Edinburgh and Dublin and won over a few more converts, but by no means did he yet have a consensus.

Dubois was buoyed by some official acclaim. The Anthropological Institute of Great Britain and Ireland made him an honorary fellow, and in 1896 he received the Prix Broca, a French prize for anthropology, for his achievement.

While all of his traveling had been necessary, Dubois wanted to settle down and to establish a base from which to continue his work. By an act of parliament, he was appointed curator of paleontology at the Teyler Museum in the Netherlands city of Haarlem. His family moved yet again and Dubois kept writing papers on and campaigning for his ape-man.

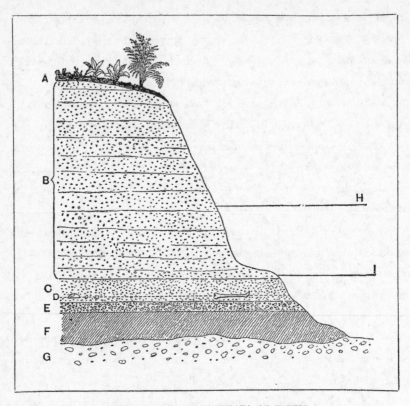

SECTION OF THE BONE STRATA AT TRINIL.

A—Vegetable soil.
B—Sand-rock.
C—Bed of lapilli-rock.
D—Level in which the four remains were found
E—Conglomerate.
F—Clay-rock.
G—Marine breccia.
H—Rainy season level of river
I—Dry season level of river.

FIGURE 4.7 *Drawing of Pithecanthropus discovery site.*
Dubois was eager to show that the fossils came from the same bed. *Drawing from E. Dubois (1896)* Journal of the Anthropological Institute of Great Britain and Ireland, 25:240–255.

In 1898 the most prominent luminaries of biology — including supporters, critics, and those still undecided — gathered at the International Congress of Zoology in Cambridge, England. Haeckel, now a senior statesman of evolutionary science, took the podium before Dubois and summoned all of his considerable eloquence to support *Pithecanthropus* and to vanquish the old-guard of Virchow and his fellow holdouts. Pronouncing Dubois the "able discoverer of *Pithecanthropus*," who had "convincingly pointed out his [the fossil's] high significance as a 'missing link'," Haeckel made such a formidable case that Dubois just sat back and enjoyed the show. Haeckel laid into Virchow in particular, enumerating point-by-point how experts disagreed with him and underscoring how Virchow had declared Neanderthals and *Pithecanthropus* to be "pathological products; indeed the sagacious pathologist at last made the incredible assertion that 'all organic variations are pathological.'" Haeckel added, "It must be remembered that for more than thirty years Virchow has regarded it as his especial duty as a scientist to oppose Darwinian theory" and to assert that "it is quite certain that man did not descend from the apes . . . not caring in the least that now almost all experts of good judgment hold the opposite conviction."

As gratifying as Haeckel's support was, the campaign was not over. Dubois kept searching for ways to convert others to his view of *Pithecanthropus*. He developed a new approach to the argument by comparing relative brain and body sizes in animals. He found a general mathematical relationship between the two in most mammals. From these data, he asked, "What size ape would have a brain almost 1,000 cc in volume?" From his studies, he calculated that it would be about a five-hundred-pound ape. But from the dimensions of the *Pithecanthropus* femur, he could tell that it would support a roughly 160-pound animal. The brain was too big to be that of an ape, and yet it was smaller than that of a modern 160-pound human. It was intermediate in size, which is just what one would expect if, as Dubois had now been arguing for nearly five years, *Pithecanthropus* was indeed an intermediate between apes and humans.

Homo erectus

It had been over a decade since Dubois, Anna, and baby Marie had boarded the ship for the East Indies. The years of criticism and debate, following the years spent searching all over Sumatra and Java, took a toll on Dubois. He was battle weary and physically worn down. Moreover, he felt he had said about all that he could about the fossils in hand; they no longer held his fascination the way they once did. As the century drew to a close, he had to satisfy himself with the knowledge that he had convinced many, but not all, of *Pithecanthropus'* place in human history, and of his own claim to a place in paleoanthropological history.

But even Dubois could not yet know the full magnitude of his feat. Following the trail that he blazed, others, many others, followed in Dubois' footsteps to Asia. Their searches demonstrated how unlikely it was that Dubois found anything, as most found little to report. Further Dutch and Prussian excavations at Trinil turned up nothing. Later, the largest land-based expedition ever mounted, a decade-long search for ancient humans in China and Mongolia led by the American Museum of Natural History, found zero human fossils (but was a smashing success in unexpected ways — see Chapter 5). It was several decades before anyone found more evidence of ancient hominids in Asia. In 1929–1930 "Peking man" (given the name *Sinanthropus pekinensis*) was described from caves in China, and in the late 1930s more *Pithecanthropus* skulls were found on Java. In 1950, these two fossil hominids were grouped together as one species, placed in the same genus as humans, and renamed *Homo erectus*. Dubois' skullcap, known as "Trinil 2," became the type specimen of *Homo erectus* — the original specimen upon which the description of a new species is established.

Dubois also could not know that his battle would typify the contentious reaction that would greet the identification of virtually every new hominid fossil and the attempt to place such discoveries in the history of human origins. New finds frequently overturn well-established notions. For example, it was not until the early 1960s, when older *Homo erectus* and other fossil hominids were found in Africa, that the focus on the "cradle of mankind" shifted fully away from Asia to Africa.

CHAPTER QUESTIONS

1. What influence did Thomas Huxley have on Eugene Dubois?

2. Why did Dubois choose to search in the Dutch East Indies for the missing link and not Africa?

3. What evidence did Dubois assemble to support his·claim that *Pithecanthropus* was an intermediate between apes and modern humans?

For more on this story, go to the *Into The Jungle* companion website at www.aw-bc.com/carroll.

FIGURE 5.1 *Roy Chapman Andrews, explorer.*
Approaching a kite's nest in the badlands of the Gobi desert of Mongolia; the ranger
hat and pistol on Roy's side were his trademarks when in the field. *Image #410988;
photo by James Shackelford, American Museum of Natural History Library.*

FIVE

Where the Dragon Laid Her Eggs

Dreams come true. Without that possibility, nature would not incite us to have them.

—John Updike

He always wanted to be an explorer.

Growing up in the late 1800s in Beloit, Wisconsin, a small industrial town on the banks of the Rock River, young Roy Chapman Andrews spent every minute he could out of doors, whatever the weather. When confined indoors, he made his mother read him *Robinson Crusoe* again and again, even though he knew every bit of the story by heart. He dreamed of living on a deserted island and fending for himself.

By age eight, Roy was passionate about nature. He wandered the woods with binoculars, a notebook, and field guides to birds. Inspired by visits to the Field Museum of Natural History in Chicago, he put together a little museum in the attic of his house in which plants, minerals, fossils, stuffed animals, and other artifacts were carefully labeled and displayed.

His parents indulged their son's burning passion. On his ninth birthday, his father gave him a small single-barrel shotgun. Young Roy took his new prize goose hunting and stalked several birds that were floating on the edge of a marsh. Crawling on his stomach through mud and water, Roy closed in and fired. Three of the geese slowly collapsed with a gentle hissing sound — he had shot someone's decoys full of holes!

The owner of the decoys, Fred Fenton, jumped out of the bushes and was not at all pleased. Roy was scared to death and took off for home sobbing. Roy's dad, however, roared with laughter over the incident. It turned out that he hated Fenton and promised to buy Roy a double-barrel shotgun. Roy's dad growled, "So you can get 'em all next time." Roy got the new weapon within a week.

Roy loved to camp, fish, and hunt and did so often on his own. One evening, after his camp dinner of bread and bacon, he curled up to sleep in the open under a great tree. During the night, he felt something wriggling in his hair and brushed at it sleepily with his hand. A cold body wrapped around his wrist and arm — a snake! Roy yelled out and was shaking with fright. It was only a harmless garter snake, but the incident haunted him for weeks, and a lifelong hatred of snakes was cemented.

Roy's love for hunting game led to an interest in taxidermy. He taught himself out of a book how to mount birds and other animals. Soon thereafter he began to mount birds and deer for local hunters. He always had plenty of work during the fall shooting season, and plenty of money each Christmas. His earnings helped him to pay his own way through the excellent local Beloit College.

Though he had a strong work ethic when it came to taxidermy, Roy was not a very diligent student. He was bored by most subjects other than science and did poorly at math. His appetite for exploration and natural history was fueled by reading accounts of the great African explorers Richard Burton, John Speke, David Livingston, and Henry Stanley, and by his many romps around the Wisconsin countryside.

Roy did take advantage of one important opportunity presented by college life — to meet professional scientists. One day he cornered a geologist from the American Museum of Natural History who had come to Beloit College to lecture. Roy explained his passion for natural history and even persuaded the visitor to drop by a local tavern to see some of Roy's mounted heads. Impressed, the geologist advised Roy to write to the museum's director about a possible job.

Roy wanted to be a naturalist and explorer so badly that no other profession had ever entered his mind. Though he had never before traveled further than the ninety miles to Chicago, on the day of his graduation he announced to his parents that he was going to New York to try to get a job at the American Museum of Natural History.

A Whale of a Start

The day after he arrived in New York, Roy made his way to the museum for an appointment with Dr. Bumpus, the director. Roy nervously answered questions before Dr. Bumpus tried to ease him down gently, explaining that there were no positions open. Roy's heart sank, but then the can-do entrepreneur rose within him. Roy blurted out, "I'm not asking for a position. I just want to work here. You have to have someone to clean the floors. Couldn't I do that?"

Dr. Bumpus responded, "A man with a college education doesn't want to clean floors!"

"No," Roy said, "Not just any floors. But the museum floors are different."

Dr. Bumpus was impressed by Roy's spark. Dr. Bumpus hired Roy and took him out to lunch. Little did Dr. Bumpus know that Roy would become the most famous museum employee, and one day hold his job.

Roy was assigned to the taxidermy department. He began each day by mopping the floors, just as he was hired to do, but he was soon given more interesting assignments. His first big opportunity came when Dr. Bumpus made Roy an assistant to a scientist who was to build a life-size model of a whale. Roy and a friend figured out that the best way to construct the model was to use wire netting and papier-mâché.

Roy had never even seen a whale up to that point. He got his chance soon thereafter when a whale carcass washed up on Long Island. Roy and his model-building friend got the assignment to bring back the whole whale skeleton. It was an enormous task to free the skeleton from the mass of meat and blubber, while the rotting carcass sank into the sand and was pounded by the surf. Soaked, frozen, and exhausted, they worked for days until they succeeded in securing this novel prize for the museum. Proud of his young protégé, Dr. Bumpus introduced Roy to many of the important visitors to the museum, from leading naturalists to such industrial tycoons of the early 1900s as Andrew Carnegie.

Roy's growing reputation concerning whales led to his first real field expedition. Little was documented about whale behavior, so Roy coaxed Dr. Bumpus into letting him accompany whaling ships operating off Vancouver. Even though he was tortured by constant seasickness, Roy was able to observe whales up close, including the birth of a calf. Roy's success on that first mission led to many more whale studies in different parts of the world over the next eight years, particularly in the Far East off Japan, Korea, and China. During this time Roy developed a strong attachment to the Orient, and expanded his reputation as a naturalist. For example, he "rediscovered" the gray whale, once thought to have been hunted to extinction. Most importantly, Roy's many expeditions, and frequent side trips, provided opportunities for adventures that fueled his appetite for exploration, and his self-confidence.

Once, when Roy had a few weeks' delay waiting for a ship out of the Philippines, he had a steamer deliver him and two Filipino assistants to an uninhabited island. On this palm-fringed paradise, surrounded by coral reefs, Roy lived the *Robinson Crusoe* experience he had dreamed of since he was a child. Perhaps it was even a bit too authentic a setting, as the ship left enough food for just five days, and a broken propeller delayed the ship's return by over two weeks.

When the food was gone, Roy didn't care — he was too happy wading in the tide pools catching fish and crabs and sleeping out under the stars. He built a snare to trap birds and lived as a castaway. When the ship finally showed up, Roy's heart was heavy, for he feared he would never again be so content as he was on "his" deserted island.

But he need not have worried, much greater adventures lay ahead.

Asia Dreaming

After eight years chasing whales around the world, Roy had enough of ocean adventures. He wanted to get onto land. But where were the greatest opportunities for discovery?

Roy was smitten with Asia, and China in particular. From the moment he entered Peking (today Beijing) he loved the colorful dress of the locals, and the stone walls, battlements, and statues that were steeped in history. He was also awed by the Great Wall. When he went to see it for the first time, he walked toward it head down so that when he looked up he would see all of its majesty at once:

> At last, there it was spreading its length like a slumbering gray serpent over the hills, into the valleys, and up the sides of precipitous peaks as far as my eyes could reach. No other sight on earth has ever stirred me as did the Great Wall of China. . . . I knew that some day I should return.

The Great Wall was built to protect Northern China from Mongol and other invaders. Roy would pass through the gates of the Great Wall again and again to explore the vast region beyond.

Roy was looking for a reason to explore Asia. He found it in the theories of the new president of the museum, Henry Fairfield Osborn. Osborn was an eminent paleontologist who had led fossil expeditions in the American West, and described and named *Tyrannosaurus rex*. Osborn believed that Asia was the home of ancient humans and the origin of much of the animal life of Europe and America. Testing these ideas required much greater knowledge of the living and fossil fauna of central Asia than was available. What better justification for an expedition could there be than to test his boss' favorite ideas? Roy proposed to mount an expedition to gather the scientific data. Osborn was, of course, quite enthusiastic.

In 1916, Roy made his first collecting foray to the Yunnan province of southeastern China. He came back with preserved specimens of twenty-one hundred mammals, one thousand birds, and many fish and reptiles, all from regions where no collector had ever been before.

While he was in Yunnan, the United States entered the First World War. Through a friend in Naval Intelligence, Roy secured an assignment to work as a spy under the cover of collecting zoological specimens for the museum. This allowed him to stay in Asia for the duration of the war. He traveled throughout China, Manchuria, Mongolia, and as far as Siberia. Roy loved the "work," as he saw "new country, new customs, new people everyday." He also filed intelligence reports on political developments and military matters. Most importantly, his travels gave him the opportunity to scout out promising places for future exploration.

He was captivated by Mongolia and the Gobi desert. He made his first trips by motorcar, a novelty at the time, and a much more comfortable mode of transport than a two-month trek on a camel. After passing through the Great Wall at Kalgan, he made his way toward Urga, the capital of Mongolia. Roy was thrilled by the colorful canyons, mountains, ravines, and the vast plains beyond. He wrote of his experience:

> Never again will I have such a feeling as Mongolia gave me. The broad sweeps of dun colored gravel merging into a vague horizon; the ancient trails once traveled by Genghis Kahn's wild raiders; the violent contrast of motor cars beside majestic camels fresh from the marching sands of the Gobi! All this thrilled me to the core. I had found my country. The one I was born to know and love.

Just after the war, in the spring of 1919, Roy made the first museum-sponsored expedition into Mongolia and into the Gobi. It was strictly a zoological collecting trip, albeit into an almost entirely unknown country. Roy's efforts produced another fifteen hundred mammal specimens for the museum, and a plan for a great expedition — the likes of which had never been imagined before.

The Big Plan

Roy went to New York to see Professor Osborn at the museum. Osborn sensed Roy had a pitch to make, and Roy let it rip. Roy told Osborn that in order to test his theories, they "should try to reconstruct the whole past history of the Central Asian plateau — its geology, fossils, past climate, and vegetation. We've got to collect its living mammals, birds, fish, reptile, insects, and plants and map the unexplored parts of the Gobi. It must be a thorough job: the biggest land expedition ever to leave the United States."

Roy had thought out the special logistical challenges the expedition presented. His experience in motorcars was crucial. He told Osborn he could cover far more ground if, instead of using camels that could cover only ten to fifteen miles a day, he used motor cars that could go one hundred miles a day and deployed a caravan of camels to act like supply ships to the fleet of cars. By sending the camels out ahead of the fleet in winter, with food and equipment, the caravan could then rendezvous with the cars deep in the desert in the summer.

Osborn peppered Roy with questions, but Roy had thought it all out in detail. Most importantly, Roy impressed upon Osborn the importance of bringing the best experts in many fields on the expedition — in geology, paleontology, mammalogy, and so forth.

Osborn was sold: "Roy, we've got to do it. This plan is scientifically sound. Moreover, it grips the imagination. Finances are the only obstacle."

Roy thought he needed two hundred fifty thousand dollars (in 1921 dollars, perhaps ten million dollars today) for a five-year expedition. He had a plan for that, too. He had rubbed shoulders with enough business titans to realize that they might support such a venture if it brought them some prestige in high society.

The Beloit taxidermist was right again.

Roy's first stop was the banker J. P. Morgan. When Roy spread his map out, Morgan listened with rapt attention. Roy laid out his whole plan in fifteen minutes. When he stopped, out of breath, Morgan blurted out, "It's a great plan; a great plan. I'll gamble with you . . . All right, I'll give you fifty thousand. Now you go out and get the rest of it." Morgan sent Roy to a fellow banker at Chase National Bank, who ponied up ten thousand dollars. Other New York elites, including John D. Rockefeller, followed suit. Roy enjoyed his adventures on Wall Street. He liked the gambling spirit of the titans of steel, oil, railroads, banks, and other industries.

The expedition soon captured the newspapers' and public's attention. Roy was deluged with thousands of letters from people, many of them teenage boys, volunteering to join the expedition. But Roy had already very carefully selected his team.

FIGURE 5.2 *The 1922 expedition team at Tsagon Nor, Mongolia.*
Second row, left to right: Morris, Colgate, Granger, Badmajapoff, Andrews, Berkey,
Larsen, Shackelford. *Top row*: Chinese technical and camp associates. *Bottom row*:
Mongolian interpreters and caravan men. *From* The New Conquest of Central Asia:
A Narrative of the Explorations of the Central Asiatic Expeditions in Mongolia
and China, 1921–1930 *by Roy Chapman Andrews (1932). The American Museum of
Natural History, New York.*

The experts Roy recruited included: Walter Granger, paleontologist and
second in command; Charles Berkey, chief geologist; Clifford Pope, herpetol-
ogist; Bayard Colgate, chief of motor transport; Frederick Morris, geologist;
and J. B. Shackelford, photographer (Figure 5.2). These individuals were very
well chosen. They got along exceedingly well and in the field their strengths
complemented one another.

With the money and the team falling into place, the major task of preparation came next. Roy put his considerable experience into the selection of the
equipment, food, and overall planning. Roy knew that it would be impossible
to obtain anything but meat in the desert so he ordered great quantities of
dried fruits and vegetables — onions, tomatoes, carrots, beets, and spinach
from America; the rest of the food was provided by a Marine Corps detach-
ment in China. He obtained Mongol tents and fur sleeping bags, as the

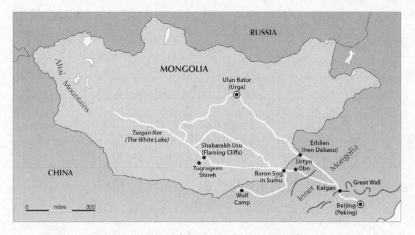

FIGURE 5.3 *Map of the Central Asiatic expeditions.*
White lines trace the various routes taken in expeditions between 1921 and 1930. Some
of the location names have changed over the years, both current and former names
are shown. *Based on map in L. Rexer and R. Klein (1995),* 125 Years of Exploration and
Discovery, *Harry N. Abrams in association with the American Museum of Natural
History, New York. Redrawn by Leanne Olds.*

nomads knew best how to cope with the desert weather. And Roy selected the
vehicles as well — three Dodge Brothers cars and two one-ton Fulton trucks.
The Standard Oil Company of New York donated three thousand gallons of
gasoline and fifty gallons of oil. Altogether, eighteen tons of equipment and
supplies were shipped to Peking. They needed seventy-five camels to carry it
all. The expedition headquarters was established in Peking, camels were pur-
chased, loaded, sent out in advance, and finally, on April 21, 1922, the team set
out in five cars from Kalgan, passed through the Great Wall, and entered the
vast territory beyond (Figure 5.3).

There were many unknowns. Foremost among them was whether they could
find any fossils. Professor Osborn had noted, and Roy also knew, that the only
fossil known to have come from the central Asian plateau was a single tooth that
had been discovered by a Russian in the late 1890s. The expedition had been
ridiculed by some as pointless; they said that the desert was a wasteland of sand
and gravel and that Roy might as well search for fossils in the Pacific Ocean. Roy
was also told it was criminal to waste the time of such eminent geologists as
Berkey and Morris in a country where the "geology was all obscured by sand."

But Roy's team would soon find out the skeptics were wrong. Oh boy, were they ever wrong.

Iren Dabasu

The car fleet quickly made its way toward the Gobi. Making and breaking camp was a fluid operation. The tents were pitched, and a fire was started within thirty minutes of selecting a spot. Each team member then took care of his particular tasks. The chief of motor transport refilled the gas tanks and thoroughly inspected each car for loose bolts or cut tires. The geologists transcribed their notes from the day's work. The photographer reloaded his film magazines and made a diary of the day's photographs. The taxidermists set traps for mammals. And, if there was a rock outcrop or exposure nearby, the paleontologists would make a quick search for fossils.

Four days into the expedition, Roy and staff set up camp at Iren Dabasu while Berkey, Morris, and Granger stopped a few miles away to scout for fossils. As Roy was enjoying a beautiful sunset over the desert, Granger and the geologists roared into camp. Granger's eyes were shining as he reached into his pockets and held out several fossils: "Well, Roy, we've done it. The stuff is here. We picked up fifty pounds of bone in an hour."

They would identify some of the teeth as from a rhinoceros, but the origins of the other mammal fragments were less certain. It didn't matter — everyone was happy and very eager to go fossil hunting at first light.

The next morning Berkey came to breakfast with his hands full of fossils; they had camped right on top of another exposure. Granger was puzzled by one leg bone — it was not mammalian. Granger went to the outcrop where Berkey had found the leg bone and exposed another bone perfectly preserved in the rock — it was a dinosaur. Berkey proclaimed, "We are standing on Cretaceous strata of the upper part of the Age of Reptiles — *the first Cretaceous strata and the first dinosaur ever discovered in Asia north of the Himalaya mountains.*"

All of Roy's optimism was validated right there. They had found what no one else had found, and were sitting on top of rich mammal and dinosaur fossil beds. There was far more to investigate at Iren Dabasu, but they didn't

FIGURE 5.4 *Expedition caravan.*
The camel caravan arriving at the Flaming Cliffs in 1925. *From* The New Conquest of
Central Asia: A Narrative of the Explorations of the Central Asiatic Expeditions in
Mongolia and China, 1921–1930 *by Roy Chapman Andrews (1932). The American
Museum of Natural History, New York.*

have the time to celebrate or to prospect for much more. They had to keep
moving to stay on schedule with the camel caravan, three hundred fifty miles
ahead of them.

Led by Merin, a remarkable Mongol, the caravan was to meet up with the
cars on April 28. And sure enough, as Roy approached the rendezvous point,
he saw the American flag flying from one of the loads. Merin had arrived at the
designated spot one hour earlier, after a thirty-eight-day trek. The sight of the
camel caravan, in single file among the rocks, was majestic (Figure 5.4).

As the expedition made its way across the desert, it repeatedly faced a stub-
born enemy — sandstorms. Roy described the power of and confusion sown by
one of the many great storms:

> Slowly I became conscious that the air was vibrating to a continuous roar,
> louder every second. Then I understood. One of the terrible desert storms
> was on the way. The shallow basin seemed to be smoking like the crater of a
> volcano — yellow "wind devils" eddied up and swirled across the plain. To

the north an ominous tawny bank advanced at horse race speed. I started back toward camp, but almost instantly a thousand shrieking storm demons were pelting my face with sand and gravel. Breathing was difficult; seeing impossible.

The wind blew for ten days, until the team's nerves were worn thin. But once the storm had passed, they again enjoyed the scenery of fantastic sand dunes, red sunsets against lavender mountains, and the fossil treasures poking out of the ground.

Shackelford, the photographer, had an uncanny knack for spotting fossils. One day, near the shore of a desert lake, he actually stumbled over a huge leg bone that had weathered out of a stream bank. The team had been told by some Mongols of "bones as large as a man's body." Now here was proof. It was the upper foreleg (humerus) of a *Baluchitherium* (or "the beast of Baluchistar," after the place in Pakistan where in 1913 it was first found). Roy and Granger later found a skull of the colossal animal — the largest land mammal that ever lived, at nearly seventeen feet tall, twenty-six feet long, and weighing fifteen tons.

Granger had his hands full with other fossils. One day alone he found 175 jaws and skulls of various carnivores, rodents, and insectivores. And nearby, he discovered a complete skeleton of a small beaked dinosaur.

As difficult as the conditions in the Gobi could be, it was actually fortunate that these fossils were in such remote locations in Mongolia. In China, fossil bones have long been interpreted as those of dragons, which are sacred symbols of power. For at least two millenia, "dragon bones" — the bones of extinct mammals and dinosaurs — have been collected, ground up into powders, and used in traditional folk medicines. Had the Gobi sites been nearer to civilization, they might have been destroyed.

✳   ✳   ✳

By September 1, it was time to leave the Gobi. The weather was turning and great flocks of birds were flying south overhead from the northern tundra. The expedition was low on water, and Roy feared being trapped in a blizzard.

FIGURE 5.5 *Protoceratops andrewsi exposed at the Flaming Cliffs.*
The small skull of this hornless dinosaur spanned just six to seven inches in length.
From The New Conquest of Central Asia: A Narrative of the Explorations of the Central
Asiatic Expeditions in Mongolia and China, 1921–1930 *by Roy Chapman Andrews
(1932).* The American Museum of Natural History, New York.

While stopped to look for a well, Shackelford went strolling. He found
himself standing on the edge of a vast red sandstone basin. Scampering down
the slope he walked toward a small rock pinnacle on top of which lay a white
fossil bone, just waiting to be plucked. It was an unfamiliar skull of a horned
dinosaur, later named *Protoceratops andrewsi* in Roy's honor (Figure 5.5).
Shackelford reported seeing other bones, so the team decided to camp.

The next day they discovered that these beautiful badlands "were paved
with white fossil bones and all represented animals unknown" to any of the
team. But they had to leave further exploration to another season. Inspired by
the fiery glow of the rocks in the late afternoon sun, Roy dubbed the spot the
"Flaming Cliffs," and the team headed back to Peking.

The scientific results of the expedition had exceeded their greatest hopes.
The team had found complete skeletons of small dinosaurs and parts of larger
ones, skulls of mastodons, rodents, carnivores, deer, giant ostrich, and rhinoceros,

as well as Cretaceous mosquitoes, butterflies, fish, and more. Shackelford had shot twenty thousand feet of film of the expedition and life in the Gobi. Almost all of their specimens were new to science. Professor Osborn sent congratulations, stating, "You have written a new chapter in the history of life upon the earth."

But the team knew they had only scratched the surface. They started preparing for the next season to return to the Flaming Cliffs.

Where the Dragon Laid Her Eggs

Exactly one year after the first expedition, the team again left Peking for the Gobi. Once camped at Iren Dabasu, Roy and another driver went back to Kalgan for more supplies. They almost didn't make it.

Approaching a deep valley where he knew Russian cars had been robbed by bandits a week earlier, Roy was wary. Sure enough, he spotted a man on horseback with a rifle. Roy took out his .38 revolver and pinged a shot nearby, just to scare the man off. But then Roy came upon four more armed horsemen. The trail was too narrow for him to turn around. So, instead, he gunned the engine and raced down the trail to scare the horses. It worked. The bandits had to hold on for dear life as their horses bolted and Roy fired off a few shots. It was, in Roy's words, "Great fun."

The team made the several hundred mile trek to the Flaming Cliffs, following the tracks their cars had made ten months earlier. They made camp in mid-afternoon, and the team members scattered to look for fossils. By nightfall, everyone had his own dinosaur skull.

On the second day, new team member George Olsen reported at lunch that he thought he had found some fossil eggs (Figure 5.6). The team gave him a good ribbing, but they were curious enough to follow Olsen back to the spot to see what he was talking about. Roy explained:

> Then our indifference suddenly evaporated. It was certain they really *were* eggs. Three of them were exposed and evidently had broken out of the sandstone ledge beside which they lay . . . We could hardly believe our eyes, but, even though we tried to account for them in every possible way as geological

FIGURE 5.6 *The first nest of dinosaur eggs.*
These seven-inch long eggs were discovered by George Olsen at the Flaming Cliffs in
1923. *From* The New Conquest of Central Asia: A Narrative of the Explorations of the
Central Asiatic Expeditions in Mongolia and China, 1921–1930 *by Roy Chapman
Andrews (1932). The American Museum of Natural History, New York.*

phenomena, there was no shadow of doubt that they really were eggs. That
they must be those of a dinosaur we felt certain. True enough, it never was
known before that dinosaurs did lay eggs . . . although hundreds of skulls and
skeletons of dinosaurs had been discovered in various parts of the world, never
had an egg been brought to light.

Dinosaur eggs! Roy later admitted of the discovery, "Nothing in the world
was further from our minds."

The eggs were not all there was to behold:

While the rest of us were on our hands and knees about the spot, Olsen scraped
away the loose rock on the summit of the ledge. To our amazement he
uncovered the skeleton of a small dinosaur lying four inches above the eggs.

It was a type of dinosaur completely new to science. Professor Osborn later
conjectured that it was an egg thief caught in the act and dubbed the new
species *Oviraptor* (egg seizer).

A few days later, five more eggs were found in a cluster, and then another
group of nine. In two eggs that had broken in half, they could plainly see the

FIGURE 5.7 *Another nest of dinosaur eggs.*
Andrews is on the right, rifle handy. Olsen is on the left. The oblong eggs are between
them. *From* The New Conquest of Central Asia: A Narrative of the Explorations of the
Central Asiatic Expeditions in Mongolia and China, 1921–1930 *by Roy Chapman
Andrews (1932). The American Museum of Natural History, New York.*

bones of dinosaur embryos inside. Twenty-five eggs in all were found that year,
and many more in subsequent years (Figure 5.7).

But the eggs did not end the treasure hunt at the Flaming Cliffs. The team
found seventy-five skulls within an area of three miles. Finding so many specimens created a problem — not enough flour. The specimens were prepared
and packed by wrapping them in cloth soaked in flour paste. In three weeks,
they had nearly exhausted all of their flour. Down to half a sack, Roy polled the
team — should they quit working or use the rest of the flour? They were unan-
imous, "Let's keep the flour for work." They had only tea and meat for food.

Flour was not the only necessity in short supply. The team also ran out of
burlap for wrapping fossils, so they had to improvise. First, they cut off all
their tent flaps; then they used their towels and wash cloths. Finally they
used their clothes — socks, trousers, shirts, underclothes, and even Roy's
pajamas. Among the many dinosaurs thus preserved Osborn would identify
several new species — including *Velociraptor*, a swift predator with large
talons, and *Tarborsaurus*, a close relative of *Tyrannosaurus rex*.

The horde from the Flaming Cliffs filled sixty empty supply boxes and gasoline tins, and weighed five tons. Buried in the massive haul was one small, as yet unrecognized gem. It was a tiny skull, barely more than an inch long, found in the same Cretaceous layer as the dinosaur eggs. Granger had labeled it as "an unidentifiable reptile." However, once unpacked and studied at the museum, it was very clear that it was not a reptile, but a mammal. It was the most complete specimen of mammal life in the Cretaceous unearthed to date — proof that mammals lived alongside dinosaurs. But with just one specimen, they had just one glimpse of early mammal life. On the next expedition finding more mammals would be a top priority.

On the Trail of Ancient Mammals

On their third visit to the Flaming Cliffs, in 1925, Roy brought along a letter that he had been carrying for Granger. It was from W. D. Matthew, curator of paleontology at the museum. In it, Matthew explained the importance of that tiny mammal skull and wrote: "Do your utmost to get some other skulls."

After Roy and Granger discussed the matter for a while Granger declared, "Well, I guess that's an order. I'd better get busy." He walked out to the Flaming Cliffs and came back with another mammal skull within an hour. Later, Granger and his assistants spent many days in the scorching sun inspecting thousands of little nodules of sandstone for more skulls. It was tedious, grueling work, but they found seven more skulls, most with lower jaws. Granger kept these hard-earned delicate specimens in his suitcase for safekeeping.

After the expedition, Roy carried them to New York and presented the treasures to Dr. Matthew. Subsequent analysis revealed that the specimens included two families of insectivores (Figure 5.8). They were some of the first "missing links" in the story of mammal evolution. The fossils revealed that before the end of the so-called "Age of Dinosaurs," mammals had already split into the two main lines of marsupial and placental forms. The Flaming Cliffs also produced a fossil representative of the "multituberculates," an ancient branch of mammals and the only major branch to have gone extinct.

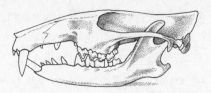

FIGURE 5.8 *An early mammal skull.*
The skull of *Zalamdalestes*, a shrew-like mammal that lived in the Cretaceous period
alongside dinosaurs, was about two inches long. *Drawn by Leanne Olds.*

Roy believed that these fossils were the most valuable finds of the entire
expedition and that Granger's intense search was "possibly the most valuable
seven days of work in the whole history of paleontology."

Snakes on a Plain!

On the return east from the Flaming Cliffs, the expedition found an abundance of geologically younger mammals at the other end of the size spectrum.
The team unearthed two fossil *Titanotheres*, including one enormous skull of
the rhinoceros-like beasts, previously known only from America. At a new
camp, no fewer than twenty-seven mammal jaws were exposed in one layer
and more lay underneath. The fossils included a strange claw-hoofed animal
and scores of a small-hoofed animal known as *Lophiodon* (related to tapirs).
Roy concluded that the region must have once swarmed with these mammals,
just as dinosaurs had also once abounded here.

The land they explored turned out to be swarming with more than fossils —
it was infested with pit vipers. During the day, three snakes were discovered
near the tents and every member of the team saw snakes while out prospecting
for fossils.

That was already far too many snakes for Roy. Then, one night, the temperature dropped to near freezing and the vipers moved into camp for warmth. All

hell broke loose. One of the motor engineers woke in the night and saw a snake near his tent door. Checking around his tent, he found snakes around each leg of his cot and one under a gasoline box. Morris the geologist cried out, "Dear God, my tent is full of snakes."

The Mongols would not kill the snakes because the team had camped in a sacred spot near a temple. The Americans had no such restraint and dispatched forty-seven vipers in total.

Everyone was on edge. Roy leaped when he saw what turned out to be a piece of rope. Granger attacked what turned out to be a pipe cleaner. No one was bitten, but after two days of finding snakes everywhere, the team packed up, left "Viper Camp," and headed for Peking.

Under a Lucky Star

When the 1925 expedition left the Flaming Cliffs, Roy noted that those few square miles of exposed desert had given them more than they had hoped for in the entire Gobi Desert — the first dinosaur eggs, a hundred skulls and skeletons of new dinosaurs, and eight Cretaceous mammal skulls. Looking at the beautiful red rock formations for what Roy thought (correctly) might be his last time, he regretted that his caravan would never again "fight its way across the long miles of desert to this treasure house of Mongolian pre-history."

Roy and his team made two more expeditions into different parts of the Gobi, in 1928 and 1930. Wars and anti-foreigner sentiment precluded expeditions in 1926, 1927, and 1929, and political upheaval eventually ended further fieldwork there altogether.

They never did find any ancient human remains, the original selling point of the expedition to its backers. It didn't matter — the expeditions' discoveries had produced enough leads to keep many scientists busy for years. The mammal fossils and dinosaurs from this region are still the subject of intense study today.

And Roy? Well, Roy was famous. The dinosaur eggs landed him on the cover of *Time* magazine. The high profile of the expedition, based in New York and backed by high society, ensured widespread press coverage of his exploits. Roy gave lectures that drew throngs, and wrote many popular magazine articles and several books about the expeditions. Honors were heaped upon him. He was given medals that had been earned by only the most intrepid explorers — Peary, Scott, Shackleton, Amundsen, and Byrd. And in 1935, Roy became the director of the museum where he once so enthusiastically swept the floors.

In New York, Roy and his wife traveled in the highest social circles and kept company with fellow explorers such as William Beebe, aviators Charles Lindbergh and Amelia Earhart, as well as various movie stars. Roy, too, was a movie star of sorts. Featured in newsreels and the papers, and usually photographed in his ranger hat with a revolver on his hip, Roy was the image of a new breed of explorer-scientist. And if all that, along with a hatred of snakes, sounds a bit like the character "Indiana Jones" of recent movies, that is perhaps not a coincidence. Indiana Jones creator George Lucas is reported to have been inspired by characters in B-movie serials of the 1940s and 1950s, which were in turn probably influenced by Roy's and others' accounts of his adventures.

In his autobiography, *Under a Lucky Star: A Lifetime of Adventure*, Roy opened the book by recalling how as a boy he "always intended to be an explorer, to work in a natural history museum, and to live out of doors." He closed the book by admitting how lucky he had been to live out his dreams, how for him: "Always there has been an adventure just around the corner — and the world is still full of corners!"

CHAPTER QUESTIONS

1. What characteristics made Roy Chapman Andrews particularly adept at exploration?

2. Why were experts in different disciplines vital to the expedition and its success?

3. What was the significance of the discovery of dinosaur eggs?

4. Why did the American Museum of Natural History scientists consider the mammal fossil so important?

For more on this story, go to the *Into The Jungle* companion website at www.aw-bc.com/carroll.

PART III

Always Expect
the Unexpected

FIGURE 6.1 *Luis and Walter Alvarez at limestone outcrop near Gubbio, Italy.*
Walter Alvarez is on the right with his right hand touching the top of the Cretaceous
limestone, at the K–T boundary. *Photograph courtesy of Ernest Orlando Lawrence
Berkeley National Laboratory.*

The Day the Mesozoic Died

The beginning of knowledge is the discovery of something we do not understand.

—Frank Herbert, *author of Dune*

Built upon the slopes of Monte Ingino in Umbria, the ancient town of Gubbio boasts many well-preserved structures that document its glorious history. Founded by the Etruscans between the second and first centuries B.C., its Roman theater, Consuls Palace, and various churches and fountains are spectacular monuments to the Roman, Medieval, and Renaissance periods. It is one of those special destinations that draws tourists to this famous part of Italy.

It was not the ancient architecture (although it certainly made the fieldwork more enjoyable), but the much older natural history preserved in the rock formations outside the city walls that brought Walter Alvarez, a young American geologist, to Gubbio in the early 1970s. Just outside Gubbio lay a geologist's dream — one of the most extensive, continuous limestone rock sequences anywhere on the planet (see Figure 6.1). The "Scaglia rossa" is the local name for the attractive pink outcrops found along the mountainsides and gorges of the area (*Scaglia* means "scale" or "flake" and refers to how the rock is easily chipped into the square blocks used for buildings, such as the Roman theatre. *Rossa* refers to the rock's pink color). The massive formation is composed of many layers that span about four hundred meters in total. Once an ancient seabed, the rocks represent some fifty million years of Earth's history.

Geologists have long used specific fossils to help identify rocks from around the world, and Walter employed this strategy in studying the formations around Gubbio. Throughout the limestone he found fossilized shells of tiny creatures called *Foraminifera*, or forams for short, a group of single-celled protists that can only be seen with a magnifying lens. But in one centimeter of clay that separated two limestone layers, he found no fossils at all. Furthermore, in the older layer below the clay, the forams were much larger than in the younger layer above the clay (Figure 6.2). Everywhere he looked around Gubbio, he found that thin layer of clay and the same distribution of forams above and below it.

Walter was puzzled. What had happened to cause such a change in the forams? How fast did it happen? How long a period of time did that thin layer without forams represent?

These questions about seemingly mundane microscopic creatures and half an inch of clay in a thirteen hundred-foot thick rock bed in Italy might appear trivial. But pursuit of answers to these questions led Walter to a truly earth-shattering discovery about one of the most important days in the history of life.

The K–T Boundary

From the distribution of fossils and other geological data, it was known that the Gubbio formation spanned parts of both the Cretaceous and Tertiary periods. The names of these and other geological time periods come from early geologists' ideas about the major intervals in Earth's history, and from some of the features that mark particular times. In one scheme, the history of life is divided into three eras — the Paleozoic ("ancient life," the first animals), the Mesozoic ("middle life," the age of dinosaurs), and the Cenozoic

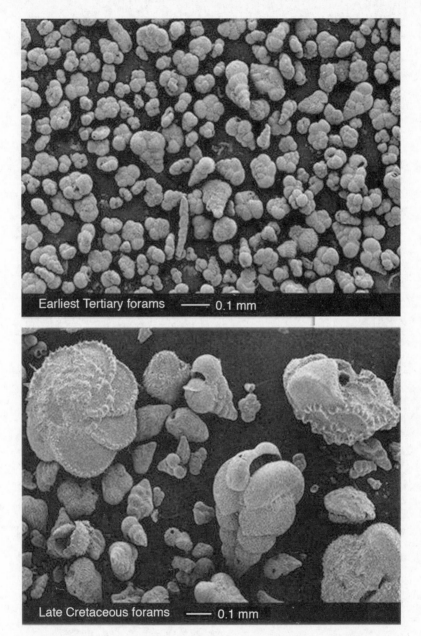

FIGURE 6.2 *Foraminifera of the Tertiary (top) and Cretaceous (bottom).*
Walter Alvarez was puzzled by the rapid, dramatic change in foram size between the
end of the Cretaceous and the beginning of the Tertiary periods, which is seen
worldwide. These specimens are from a different location (not Gubbio).
Images courtesy of Brian Huber, copyright Smithsonian Museum of Natural History.

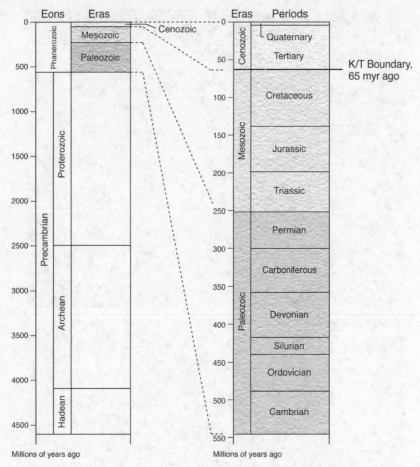

FIGURE 6.3 *Geologic time scale.*
Figure by Leanne Olds.

("recent life," the age of mammals) (Figure 6.3). The Cretaceous period, named after characteristic chalky deposits, forms the last third of the Mesozoic era. The Tertiary period began at the end of the Cretaceous period sixty-five million years ago and ended at the beginning of the Quaternary period 1.6–1.8 million years ago.

Walter and his colleague, Bill Lowrie, spent several years studying the Gubbio formation, sampling up from the Tertiary and down through the Cretaceous. They were first interested in trying to decipher geological history.

FIGURE 6.4 *The K–T boundary at Gubbio.*
The white Cretaceous limestone is separated from the reddish Tertiary limestone
by a thin clay layer (marked with coin). *Photo courtesy of Frank Schönian,
Museum of Natural History, Berlin.*

Their strategy was to correlate reversals in the Earth's magnetic field, which occur at irregular intervals and leave a distinct signature on rock grains, with the fossil record of forams. They learned to figure out where they were in the rock formation by the forams characteristic of certain deposits, and by learning to recognize the boundary between the Cretaceous and Tertiary rocks. That boundary was always right where the dramatic reduction in foram size occurred. The rocks below were Cretaceous and the rocks above were Tertiary, and the thin layer of clay was in the gap between (Figure 6.4). This boundary is referred to as the K–T boundary (K is the traditional abbreviation for the Cretaceous, T for the Tertiary).

When another geologist, Al Fischer, pointed out to Walter that the K–T boundary was about the same age as the most famous extinction of all — that of the dinosaurs — Walter became even more interested in those little forams and the K–T boundary.

Walter was relatively new to academic geology. After he received his Ph.D., he had worked for the exploration arm of a multinational oil company in Libya, until Colonel Qaddafi expelled all of the Americans from the country. Walter's work on magnetic reversals had gone well, but he realized that the abrupt change in the Gubbio forams and the K–T extinction presented a much bigger mystery, which he became determined to solve.

One of the first questions Walter wanted to answer, naturally, was how long did it take for that thin clay layer to form? To answer this he would need some help. It is very common for children to get help from their parents with their science projects, however, it is extremely unusual, as it was in Walter's case, that the "child" is in his late 30s. But few children of any age had a dad like Walter's.

From A-Bombs to Cosmic Rays

Walter's father, Luis Alvarez, knew very little about geology or paleontology, but he knew a *lot* about physics. He was a central figure in the birth and growth of the science of nuclear physics. He received his Ph.D. in physics in 1936 from the University of Chicago and worked at Berkeley under Ernest Lawrence, the recipient of the 1939 Nobel Prize in Physics for his invention of the cyclotron.

Luis' early work in physics was interrupted by the onset of World War II. During the first years of the war, he worked on the development of radar and systems that would help airplanes land safely in poor visibility. He received the Collier Trophy, the highest honor in aviation, for developing the Ground Controlled Approach system (GCA) for bad weather landings.

In the middle of the war, he was recruited into the Manhattan Project, the top secret national effort to develop atomic weapons. Alvarez and his student, Lawrence Johnston, designed the detonators for the explosives that would be used on two bombs. Robert Oppenheimer, the director of the Manhattan Project, then put him in charge of measuring the energy released by the bombs. Luis was one of the very few to witness the first two atomic blasts. He flew as a scientific eyewitness to the first test of the atomic bomb in the New Mexico desert and then shortly thereafter to the bomb dropped on Hiroshima, Japan.

After the war, Luis returned to physics research. He developed the use of large liquid hydrogen bubble chambers for tracking the behavior of particles.

In 1968, Luis received the Nobel Prize in Physics for his work in particle physics.

That would seem to have been a nice capstone to an illustrious career, but several years later his son Walter moved to Berkeley, where Luis had worked for many years, to join the university's geology department. This gave father and son the chance to talk often about science. One day, Walter gave his dad a small, polished cross-section of Gubbio K–T boundary rock and explained the mystery within it. Luis, then in his late 60s, was hooked and started thinking about how to help Walter solve that mystery. They started brainstorming about how to measure the rates of deposition around the K–T boundary. They needed some kind of atomic timekeeper.

Luis, an expert on radioactivity and decay, first suggested that they measure the abundance of beryllium-10 in the K–T clay. This isotope is constantly created in the atmosphere by the action of cosmic rays on oxygen. The more time needed to deposit the clay, the more beryllium-10 would be present. Luis put Walter in touch with a physicist who knew how to do the measurements. But just as Walter was set to work, he learned that the published half-life of beryllium-10 was wrong. The actual half-life was shorter, and too little beryllium-10 would be left after sixty-five million years to measure it.

Fortunately, Luis had another idea.

Space Dust

Luis remembered that meteorites are ten thousand times richer in elements from the platinum group (ruthenium, rhodium, palladium, osmium, iridium, and platinum) than is the earth's crust. He figured that the rain of dust from outer space should be falling, on average, at a constant rate. Therefore, by measuring the amount of space dust (platinum elements) in rock samples, one could calculate how long they had taken to form.

Platinum elements are not abundant, but they are measurable. Walter figured that if the clay bed had been deposited over a few thousand years, it would contain a detectable amount of platinum group material, but if it had been deposited more quickly, it would be free of these elements.

Luis decided that iridium, not platinum itself, was the best element to measure because it was more easily detected. He also knew the physicists that could do the measurements, the two nuclear chemists Frank Asaro and Helen Michel at the Berkeley Radiation Laboratory.

Walter gave Asaro a set of samples from across the Gubbio K–T boundary. For months, Walter heard nothing back. The analytical techniques Asaro was using were slow, his equipment was not working, and he had other projects to work on.

Nine months later Walter got a call from his dad. Asaro wanted to show them his results. They had expected iridium levels on the order of 0.1 parts per billion of sample. Asaro found three parts per billion of iridium in the clay layer, about *thirty times* more than the level found in other layers of the rock bed (Figure 6.5).

Why did that thin layer have so much iridium?

Before Walter, Luis, and Asaro got too carried away with speculation, it was important to know if the high level of iridium was an anomaly of rocks around Gubbio, or a more widespread phenomenon. Walter went looking for another exposed K–T boundary site that they could sample. He found such a place at Stevns Klint, south of Copenhagen, Denmark. Walter visited the clay bed there and could see right off that "something unpleasant had happened to the Danish sea bottom" when the clay was deposited. The cliff face was almost entirely made of white chalk, full of all kinds of fossils. But the thin K–T clay bed was black, stunk of sulfur, and had only fish bones in it. Walter deduced that during the time this "fish clay" was deposited, the sea was an oxygen-starved graveyard. He collected samples and delivered them to Frank Asaro.

In the Danish fish clay, iridium levels were 160 times typical levels.

Something very unusual, and very bad, had happened at the K–T boundary. The forams, the clay, the iridium, the dinosaurs were all signs — but of what?

It Came from Outer Space

The Alvarezes concluded right away that the iridium must have been of extra-terrestrial origin. They thought of a supernova, the explosion of a star that

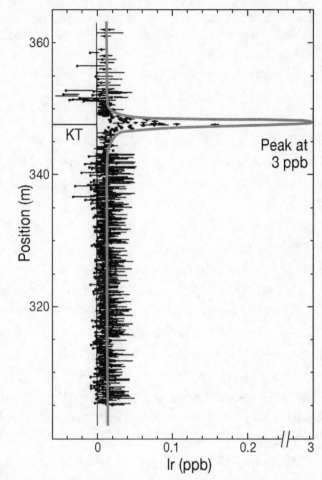

FIGURE 6.5 *The iridium anomaly.*
The levels of iridium across the Gubbio formation are plotted. Note the huge spike at
the K–T boundary. *The figure is redrawn from Alvarez et al.*, Science 250:1700–1702
(1990) by Leanne Olds.

could shower the earth with its elemental guts. The idea had been kicked
around before in paleontological and astrophysics circles.

Luis knew that heavy elements are produced in stellar explosions so, if that
idea was right, then there would be other elements besides iridium in unusual
amounts in the boundary clay. The key isotope to measure was plutonium-244
with a half-life of seventy-five million years. It would be still present in the clay

layer, but decayed in ordinary earth rocks. Rigorous testing proved there was no elevated level of plutonium. Everyone was at first disappointed, but the sleuthing continued.

Luis kept thinking of some kind of scenario that could account for a world-wide die-off. He thought that maybe the solar system had passed through a gas cloud, that the sun had become a nova, or that the iridium could have come from Jupiter. None of these ideas held up. An astronomy colleague at Berkeley, Chris McKee, suggested that an asteroid could have hit the earth. Luis at first thought that an asteroid impact would only create a tidal wave, and he could not see how even a giant tidal wave could kill the dinosaurs in Montana, Mongolia, and elsewhere.

Then he started to think about the volcanic explosion on the island of Krakatoa, in 1883. He recalled that miles of rock had been blasted into the atmosphere and that fine dust particles had circled the globe and stayed aloft for two years or more. Luis also knew from nuclear bomb tests that radioactive material mixed rapidly between hemispheres. Maybe a huge amount of dust from a large impact could turn day into night for a few years, cooling the planet and shutting down photosynthesis?

If so, how big an asteroid would it have been?

From the iridium measurements in the clay, the concentration of iridium in so-called chondritic meteorites (stony chunks from asteroids), and the surface area of Earth, Luis calculated the mass of the asteroid to be about three hundred billion metric tons. He then used various methods to infer that the asteroid would have had a diameter of 10 ± 4 kilometers.

That diameter might not seem enormous with respect to the thirteen thousand kilometer (eight thousand mile) diameter of Earth, but now consider the energy of the impact. Such an asteroid would enter the atmosphere traveling at about twenty-five kilometers per second — over fifty thousand miles per hour. It would punch a hole in the atmosphere ten kilometers across and hit the planet with the energy of 10^8 *megatons* of TNT. (The largest atomic bomb ever exploded released the equivalent of about one megaton — the asteroid was one hundred million times more powerful.) With that energy, the impact crater would be about two hundred kilometers across and forty kilometers deep, and immense amounts of material would be ejected into the atmosphere and beyond from the impact.

The team had their foram- and dinosaur-killing scenario.

Hell on Earth

The asteroid crossed the atmosphere in about one second, heating the air in front of it to several times the temperature of the sun. On impact, the asteroid vaporized, an enormous fireball erupted out into space, and rock particles were launched as far as halfway to the moon. Huge shock waves passed through the bedrock, then curved back up to the surface and shot melted blobs of bedrock out to the edge of the atmosphere and beyond. A second fireball erupted from the pressure on the shocked limestone bedrock. For a radius of a few hundred kilometers from ground zero, life was annihilated. Further away, matter ejected into space fell back to earth at high speeds — like trillions of meteors — heated up on reentry, and ignited forest fires across continents. Tsunamis, landslides, and earthquakes further ripped apart landscapes nearer to the impact.

Elsewhere in the world, death came a bit more slowly.

The debris and soot in the atmosphere blocked out the sun and the darkness may have lasted for months. This shut down photosynthesis and halted food chains at their base. Animals at successively higher levels of the food chain also succumbed. The K–T boundary marks more than the end of the dinosaurs, it was also the end of belemnites and ammonites (two prominent groups of molluscs), as well as marine reptiles. Paleontologists estimate that 50 percent of all marine genera, and perhaps 80–90 percent of all marine species went extinct. On land, nothing larger than twenty-five kilograms in body size survived.

It was the end of the Mesozoic world.

Where is the Hole?

Luis, Walter, Asaro, and Michel put together the whole story — the Gubbio forams, the iridium anomaly, the asteroid theory, the killing scenario — in a single paper published in the journal *Science* in June 1980. It is a remarkable, bold synthesis across different scientific fields, perhaps unmatched in scope by any other single paper in the modern scientific literature.

The researchers were concerned that the scientific community was not well prepared to accept their theory. They had done as much as they could to test their case, even performing an iridium analysis on a New Zealand K–T bed just to double-check their findings. The sample showed a twenty-fold iridium spike, confirming that the phenomenon was global. They had good reason to be worried. For the previous 150 years, since the beginning of modern geology, the emphasis had been on the power of gradual change. The science of geology had supplanted biblical stories of catastrophes. The idea of a catastrophic event on Earth was not just disturbing, it was considered unscientific. Until the asteroid paper, explanations for the disappearance of the dinosaurs usually invoked gradual changes in climate or in the food chain to which the animals could not adapt.

Some geologists scoffed at the catastrophe scenario and some paleontologists were not at all persuaded by the asteroid theory. Some pointed out that the highest dinosaur bone in the fossil record at the time was three meters below the K–T boundary. Perhaps the dinosaurs were already gone when the asteroid hit? Other paleontologists rebutted that because dinosaur bones are so scarce one should not expect to find them right up against the boundary. Rather, they argued the rich fossil record of forams and other creatures is the more revealing record, and forams and ammonites do persist right up to the K–T boundary.

Of course, there was a somewhat larger problem that begged explanation: Where on Earth was that huge crater? To both the skeptics and proponents this was an obvious weakness of the theory, and so the hunt was on to find the impact zone, if it existed.

At the time, there were only three known craters on Earth one hundred kilometers or more in size. None were the right age. If the asteroid had hit the ocean, which after all covers more than two-thirds of the planet's surface, then searchers might be out of luck. The deep ocean was not well mapped and a substantial part of the pre-Tertiary ocean floor had been swallowed up into the deep Earth in the continual movement of tectonic plates.

In the decade following the proposal of the asteroid theory many clues and trails were pursued, often to dead ends. As the failures mounted Walter began to believe that the impact had in fact been in an ocean.

Then a promising clue emerged from a riverbed in Texas. The Brazos River empties into the Gulf of Mexico. The sandy bed of the river is right at the K–T boundary. When examined closely by geologists familiar with the pattern of deposits left by tsunamis, the sandy bed was found to have features that could only be accounted for by a giant tsunami, perhaps fifty to one hundred meters high.

Many scientists were on the hunt for the impact site. Alan Hildebrand, a graduate student at the University of Arizona, was one of the most tenacious. Hildebrand concluded that the Brazos River tsunami bed was a crucial hint to the crater's location — that it was in the Gulf of Mexico or the Caribbean. He looked at available maps to see if there might be any candidate craters around. He found some rounded features on maps of the sea floor north of Colombia. He also learned of some circular-shaped "gravity anomalies," places where the concentration of mass varies, on the coast of Mexico's Yucatan Peninsula.

Hildebrand searched for any other hints that he was on the right track. Two signatures of impact events are glass spherules, called *tektites*, that form under the tremendous temperatures and pressure of impact (Figure 6.6) and microscopic "shocked" quartz grains. Hildebrand noticed a report of tektites in late Cretaceous rocks from a site on Haiti. When he visited the lab that had made the report, he recognized the material as impact tektites. He then went to Haiti and discovered that the deposits there included the largest tektites and shocked quartz grains ever found. He and his advisor William Boynton surmised that the impact site was within one thousand kilometers of Haiti.

After Hildebrand and Boynton presented their findings at a conference, they were contacted by Carlos Byars, a reporter for the *Houston Chronicle*. Byars told Hildebrand that geologists working for the state-owned Mexican oil company PEMEX might have discovered the crater many years earlier. Glen Penfield and Antonio Camargo had studied the circular gravity anomalies in the Yucatan. PEMEX would not allow them to release company data, but Penfield and Camargo did suggest at a conference in 1981, just a year after the

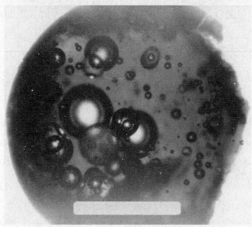

FIGURE 6.6 *Tektites.*
Tektites from Dogie Creek, Wyoming (*top*) and Beloc, Haiti (*bottom*). Note the
bubbles within the glassy sphere. These formed in the vacuum of space as the
particles were ejected out of the atmosphere. *Top photo courtesy of Geological Survey
of Canada and Alan Hildebrand. Bottom figure from J. Smit,* Annual Review of Earth
and Planetary Science *(1999) 27:75–113, used with permission.*

Alvarezes' asteroid proposal, that the feature they mapped might be the crater.
Penfield even wrote to Walter Alvarez with that suggestion.

In 1991, Hildebrand, Boynton, Penfield, Camargo, and colleagues for-
mally proposed that the 180-kilometer diameter crater (almost exactly the

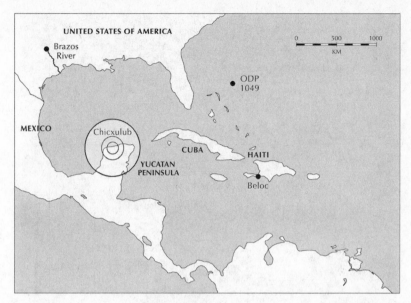

FIGURE 6.7 *Location of Chicxulub crater and key impact evidence sites.*
The map shows locations of various impact evidence — the tsunami bed in the Brazos
River, tektites in Haiti, the ocean drilling site 1049 (see Figure 6.9), and the crater and
surrounding ejected material on the Yucatan peninsula. *Drawn by Leanne Olds.*

size predicted by the Alvarez team) one-half mile below the village of
Chicxulub [Cheech-zhoo-loob] on the Yucatan Peninsula was the long-
sought K–T impact crater (Figure 6.7).

There were still crucial tests to be done to determine if Chicxulub was truly
the "smoking gun." The first issue was the age of the rock. Because the crater
was buried, determining its age was no easy task. The best approach would be to
test the core rock samples from the wells drilled by PEMEX in the region
decades earlier. At first, it was feared that all of the core samples had been
destroyed in a warehouse fire. The core samples were eventually located, and
the rock that was melted by the impact of the asteroid could be dated by a number of laboratories. The results were spectacular. One lab obtained a figure of
64.98 + 0.05 million years, another a value of 65.2 + 0.4 million years. Right on
the button — the melted rock was the same age as the K–T boundary.

The Haitian tektites were also dated to this age, as was a deposit of material
ejected from the asteroid impact. Detailed chemical analysis showed that the

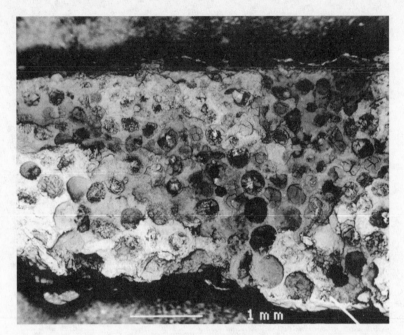

FIGURE 6.8 *A K–T spherule layer.*
An exquisitely well-preserved site near Tbilisi, Republic of Georgia. A magnified
view of the deposit reveals a graded layer of spherules (smaller particles at the
top, larger at the bottom) ejected from the impact that is also highly enriched
in iridium (86 ppb). *Image from J. Smit*, Ann. Rev. Earth Planet.
Sci. *(1999) 27:75–113, used with permission.*

Chicxulub melt rock contained high levels of iridium and that it and the
Haitian tektites came from the same source. Furthermore, the Haitian tektites
had extremely low water content and the gas pressure inside was nearly zero,
indicating that the glass solidified while in ballistic flight outside the atmosphere.

Within a little more than a decade, what had at first seemed to be a radical,
and to some, an outlandish idea, had been supported by all sorts of indirect evidence, and then ultimately confirmed by direct evidence. Geologists subsequently identified ejected material that covers most of the Yucatan and is
deposited at more than one hundred K–T boundary sites around the world
(Figures 6.8 and 6.9).

The identification of the huge crater, while a great advance for the asteroid
theory, was bittersweet for Walter. Luis Alvarez had passed away in 1988, just
before the Chicxulub discovery.

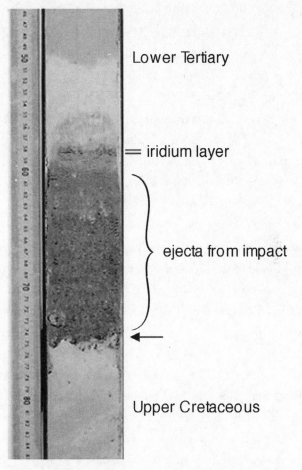

Lower Tertiary

iridium layer

ejecta from impact

Upper Cretaceous

FIGURE 6.9 *The K–T impact documented in an ocean core sample.*
This core sample, drilled at a site about five hundred kilometers east of
Florida (Ocean Drilling Project site 1049), beautifully depicts the K–T event.
The arrow marks the point in the geological column when debris from
the impact was beginning to be deposited. Note the very large layer of ejected
material on top of which the iridium-containing layer settled.
Image courtesy of Integrated Ocean Drilling Program.

Out of the Ashes

Luis' ideas, however, have continued to influence geology and the study of
the K–T boundary. In 2001, a team of scientists used Luis' idea of measuring
space dust particle accumulation to estimate the duration of the K–T

boundary period, the problem that Walter had first asked his dad for help with. The researchers used an isotope of helium-3 instead of beryllium or iridium, as a timekeeper. They estimated that the K–T-boundary clay was deposited in roughly ten thousand years. They also examined one particularly well-resolved K–T boundary site in Tunisia, Africa, where a very thin (2–3 mm thick) layer is present at the base of the boundary clay. This layer contains shocked quartz grains and other residue of the material ejected from the impact zone that spread throughout the atmosphere and then settled back to Earth. This thin fallout layer was estimated to be deposited over a course of about sixty years.

These findings suggest that it took about ten thousand years for ocean food chains and ecosystems to recover to the point where the microscopic fauna were repopulated to the levels present prior to the impact. But many of the larger ocean and land animals never recovered.

Instead, in the wake of the destruction of the Mesozoic world, a new era emerged as the Cenozoic, the age of mammals. This group of generally small species, such as those found on Roy Chapman Andrews' Asiatic expedition (see Chapter 5), took advantage of the niches vacated in the demise of the Cretaceous. Mammals evolved rapidly into many various-sized species, including large herbivores and carnivores. Within ten million years, forms representing most modern orders appear in the Cenozoic fossil record. The end of the dinosaurs was certainly the mammals' gain.

Other Smoking Guns?

The lessons learned from Chicxulub have spawned intense interest in identifying other impacts that may have affected life on Earth. The K–T extinction was not the largest on record. That dubious honor belongs to the Permian–Triassic transition during which perhaps as many as 90 percent of species living in the late Permian disappeared in an interval of less than two hundred thousand years about 251 million years ago. While there are many ideas being investigated for this mass extinction, two large candidate craters were recently proposed as evidence of a Permian impact: the two hundred kilometer Bedout crater off the northwest coast of Australia and an even larger crater buried under the ice of Wilkes Land, Antarctica.

The Chicxulub discovery has also inspired astronomers to scan the sky for other potential inbound asteroids. There are thousands of asteroids close to Earth's orbit. On March 23, 1989, a one thousand-foot diameter asteroid missed the earth by four hundred thousand miles, passing through the exact position that the planet was in just six hours earlier. Had that asteroid struck, it would have a caused a greater than one thousand megaton explosion, the largest in recorded history.

We now understand that the history of life on Earth has not at all been the orderly, gradual process envisioned by generations of geologists since Charles Lyell and Charles Darwin. There are more than 170 confirmed impact sites of various sizes on our planet — and certainly more to come.

CHAPTER QUESTIONS

1. The extinction of the dinosaurs, ammonites, and other creatures at the end of the Cretaceous was well known for many decades prior to the Alvarezes' discovery. Why did Walter Alvarez, a geologist, and Luis Alvarez, a nuclear physicist, crack a mystery that paleontologists and biologists had not?

2. What were the pieces of evidence that led to the identification of Chicxulub as the asteroid impact site?

3. What might be the local and global effects of future smaller or similar-sized asteroid impacts?

For more on this story, go to the *Into The Jungle* companion website at www.aw-bc.com/carroll.

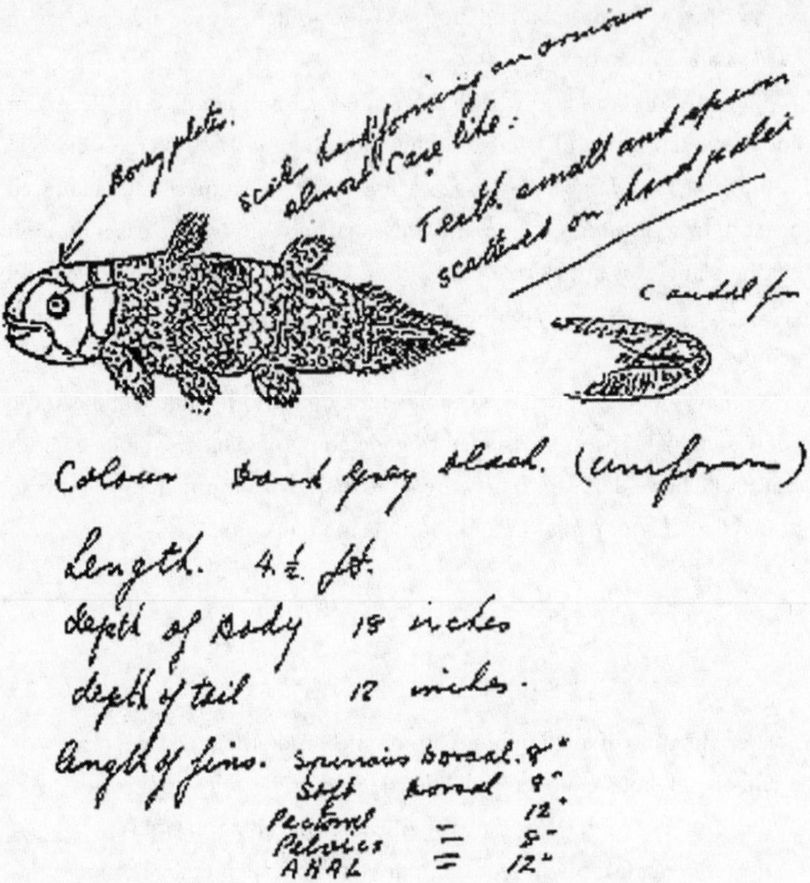

FIGURE 7.1 *Latimer's sketch and description of her mysterious fish.*
This is a copy of the original sketch Marjorie Courtenay-Latimer sent
to J. L. B. Smith on December 23, 1938. *Figure composed by Jamie Carroll.*

SEVEN

Miss Latimer's Extraordinary Fish

"Our brightest blazes of gladness are commonly kindled by unexpected sparks."

—Samuel Johnson (1709–1784)

December 22, 1938 was a sweltering day across South Africa. Marjorie Courtenay-Latimer was busy trying to put together a dinosaur skeleton she had excavated before the museum closed for the holidays.

Mid-morning, Marjorie received a phone call from the manager of the local trawler fleet — the *Nerine* had arrived at the docks and it might have caught some fish she might like for her collection. Marjorie was the first curator of the small East London Natural History Museum and had built up the small museum's collection of fossils, birds, fish, and other animals from scratch. "Oh dear," she thought, "so much still to do." Then she thought of how good Captain Goosen and the men of the trawler company had been to her over the years. Goosen had transported crates of specimens for her after an expedition to Bird Island, and he and his men collected live fish for the museum's aquarium. The least she could do was go wish the crew Happy Christmas.

Marjorie put aside her dinosaur and went down to the dock. All of the crew was gone except one. She boarded the fishing trawler and eyed the fragrant mound of fish, sponges, sharks, and other creatures now baking in the heat of the sun. There was nothing special, and she was starting to leave when she spotted it. As she pulled away a pile of carcasses she saw, "the most beautiful fish I had ever seen. . . . It was 5 feet long and a pale mauve-blue with iridescent markings."

It was also unlike any other fish she had ever seen. It was covered in hard scales, had four limb-like fins, and a strange, puppy-dog tail. She knew it had to be preserved. The fish weighed 127 pounds, so getting the dead and decomposing creature back to the museum was no small matter. It took quite a bit of persuasion to get a taxi driver to allow it into his trunk.

Once back at her post, she showed off her catch to the museum's chairman, Dr. Bruce-Bays, a physician. He was a sarcastic man who was fond of calling Marjorie "Mistress Madge," and promptly dismissed the fish as an ordinary rock cod.

Marjorie ignored him and tried to find some way to preserve the hulk that was lying on her examination table. Always resourceful, if not unconventional, her first thought was the town's mortuary. So she put the fish on a cart and wheeled it down the street, much to the disapproval of East London's pedestrians out shopping on that hot day. The man in charge of the morgue was even more rattled and refused her on the spot. Marjorie appealed, "Ah, well, they're all asleep, and it's such a beautiful fish." He did not budge.

She next tried the cold storage company, but the man there was afraid the spoiling fish would ruin the food inside, so that too was out. She then thought of the taxidermist, who had known her since she was little and trained her in preserving and mounting animals. He had also never seen a fish like Marjorie's, and he helped her wrap it in formalin to buy some time.

But what was *it*? Virtually self-taught in natural history, Marjorie consulted all of the fish books that she had. None of her references showed anything like the new blue fish. She decided to call Dr. J. L. B. Smith, a chemistry lecturer and amateur ichthyologist at Rhodes University one hundred miles away. Smith acted as honorary curator for small museums along the southern coast of the country. When she could not reach Smith by phone, she wrote him a brief letter the next day, enclosing a description and a drawing of the fish (see Figure 7.1).

Then Marjorie waited. Day after day went by without a reply from Smith. The fish started to decay and something more had to be done than her makeshift wrapping in preservative. She asked the taxidermist to preserve the scales and whatever else he could, but the innards would have to be tossed.

Smith did not see Marjorie's letter for eleven days, not until after the New Year. He was recovering from an illness, and when he finally had the chance to open the letter he was bewildered. He did not recognize the fish as South African, or from anywhere else for that matter. But then, Smith later recalled, "A bomb seemed to burst in my brain, and beyond that sketch and the paper of the letter I was looking at a series of fishy creatures that flashed up as on a screen, fishes no longer here, fishes that had lived in dim past ages gone, and of which only fragmentary remains in rocks are known."

Smith realized the letter was already eleven days old and worried he might be too late, he immediately wired Marjorie: "MOST IMPORTANT PRE-SERVE SKELETON AND GILLS FISH DESCRIBED."

Smith was stirred up by a possibility that his brain kept telling him was impossible. But Marjorie's sketch, and then some scales he received later, told him that this fish was a coelacanth, a member of a group of fish with paired fins thought to be closely related to the first four-legged vertebrates, *and thought to have been extinct since the end of the Cretaceous period, sixty-five million years ago.*

A month later, Smith went to East London to see the fish in person, which removed all of his doubts, as well as inhibitions. "Although I had come pre-pared, that first sight hit me like a white hot blast and made me feel shaky . . . I forgot everything else and just looked and looked, and then almost fearfully went close up and touched and stroked [it]."

It was time to tell the world.

A Living Fossil

J. L. B. Smith was perhaps nearly as peculiar a creature as the fish he stared at. He had a photographic memory, which no doubt helped him make the connection between Marjorie's rough sketch and fossils he had seen in the scientific literature. He could read sixteen languages and speak eight. He was absolutely devoted to Work. (Smith was fond of capitalizing the word when writing grade reports of his students.) He followed a strict, spartan diet, never eating meat with vegetables or bread with butter or cheese. He left no room in his life for luxuries or frills.

His passion was fish.

Smith certainly thought he was the right person for the job. He had long been convinced that he was destined to discover "some quite outrageous creature" and now here was the coelacanth fulfilling his premonition. The fish was sent, under police guard, to Smith's house for study. Smith issued strict instructions to his family on fish security: It was never to be left alone in the house, and in case of fire, the fish was to be the first saved. He spent every minute of the next several weeks studying every detail of the fish.

Smith was sure that the world of zoology was going to be electrified by the discovery. He also knew that, working far off the map of the world's scientific centers, he would have to get the story right the first time.

What was all the excitement about? Was the coelacanth "just" a strange fish that was mistakenly thought to be extinct? No, it was much more than that.

Biologists and paleontologists had long recognized three distinct groups of fish: sharks and rays; the so-called "ray-finned" fish, which included most familiar species (salmon, cod, eels, you name it); and the "lobe-finned" fish, which included lungfishes and, up until Marjorie's find, the extinct coelacanths.

The earliest coelacanth fossils date to the Middle Devonian period, about 380 million years ago. About 125 fossil species have been found that date up to the end of the Cretaceous period, sixty-five million years ago. The group was thought to have died out at the same time as the dinosaurs. What helped Smith recognize Marjorie's fish as a coelacanth from just her sketch was that

FIGURE 7.2 A *living fossil.*
Compare Latimer's sketch with a fossil coelacanth from 250 million years ago.
Note the great similarity in body forms. *Figure composed by Jamie Carroll.*

the fish so closely resembled its extinct relatives (Figure 7.2). The body form of
the coelacanth had changed relatively little over 380 million years, earning the
new coelacanth the moniker of "a living fossil." (This does not mean that the
living species was 380 million years old, but that it was descended from a series
of ancestors extending back to this period.)

The coelacanth body parts of foremost interest, at the time of Marjorie's
find and now, are its limb-like fins. The fundamental distinction between the
"ray-finned" and "lobe-finned" fish is in how the fin attaches to the girdle
of the main body. In ray-finned fishes, there are several bones that attach
(Figure 7.3). But in lobe-finned fishes, the paired fins are attached via single
bones, and these bones are in the same position as the humerus and femur
of the limbs of four-legged vertebrates (tetrapods), including humans (see
Figure 7.3). The way this single bone attaches allows for greater rotation of
the limbs. During the late Devonian (360–380 million years ago) lobed fins
evolved into the walking limbs of land vertebrates. Paired fins and certain
bones within them are thus said to be homologous to tetrapod limbs and

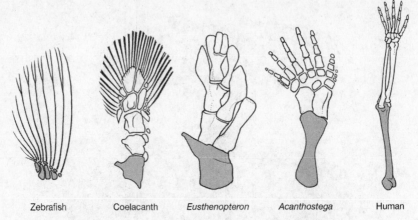

| Zebrafish | Coelacanth | Eusthenopteron | Acanthostega | Human |

FIGURE 7.3 *Lobe-fins and the evolution of limbs.*
A pectoral fin of a zebrafish (a ray-finned fish); a pectoral fin of the living coelacanth
Latimería chalumnae; pectoral fins of the extinct lobe-fin *Eusthenopteron;* the
hindlimb of the tetrapod *Acanthostega;* and a human arm. The homologous parts
of the fins and limbs are shaded. *Drawn by Leanne Olds.*

bones — they are variations of structures that were present in a common ancestor. The discovery of the living coelacanth gave scientists, and the world, the opportunity to see in the flesh a fairly good approximation of what our fish ancestors looked like.

Smith completed his first report on the fish and had claim to its naming rights. He called it *Latimeria chalumnae,* in honor of Marjorie Courtenay-Latimer and the Chalumna River near where it was caught. He submitted the report and a photograph of the fish to *Nature,* then and now a most prestigious scientific journal.

The press went crazy. They seized on the notion of a "living fossil" and likened the discovery of the coelacanth to that of finding a living dinosaur. News articles appeared throughout the world (Figure 7.4).

Marjorie's little East London Museum was famous. When Smith returned the fish to the museum, large crowds wanted to see the new wonder. Every museum coveted a specimen, perhaps too much so. One day Marjorie was given a letter to type by Dr. Bruce-Bays; it was an offer to sell the specimen to the British Museum of Natural History in London. Marjorie refused to type it,

FIGURE 7.4 *Marjorie with coelacanth.*
This picture circulated throughout the world.

and said she would resign rather than do so. Bruce-Bays relented, and the specimen is still on display today at the East London Museum in South Africa.

The museums were not the only ones to covet more specimens. Scientists worldwide wanted their own coelacanth to study, and Smith wanted at least another specimen in order to study the internal organs, which had been removed and disposed of before he could examine them. Many questions still remained. Where did the coelacanth live? Was it a deep ocean creature, as some believed, or a reef- and rock-dweller, as Smith surmised?

He, and everyone else, would wait a long time to see another coelacanth.

Wanted Dead or Alive

Shortly after the coelacanth circus, the outbreak of World War II preempted any expeditions to find more fish. But the passage of the years did nothing but intensify Smith's eagerness to get another specimen. And, over the years he had gained a great deal of prestige and influence. He convinced the South African government to put up the funds for a one hundred pound reward for anyone who caught a coelacanth and turned it over for scientific study. Marjorie arranged a special coelacanth exhibition at her museum and gave the proceeds to Smith so he could have reward posters printed in French and Portuguese and distributed all over the East African coast.

On Christmas Eve 1952, Smith received a telegram from Dzaoudzi, a small islet in the Comoros Islands. The telegram was sent by Captain Eric Hunt, who wrote that he had a five foot coelacanth and had injected it with formalin.

Could it be? Fourteen years had now passed since Marjorie's electrifying letter. Did Hunt really have the right fish?

There was only one way to find out and that was to fly to the Comoros. But it was Christmas Eve, and the Comoros was a French territory more than fifteen hundred miles away. How would Smith get there and secure Captain Hunt's fish for South Africa?

Smith tried to reach various scientific and government officials, even the ministers of economic and internal affairs. They were all away. Precious time was slipping by. Smith received another cable from Hunt saying that local authorities were going to claim the specimen unless Smith showed up in person. Smith became, in his words, "Possessed." He decided to contact someone with clout — the prime minister, Dr. D. F. Malan. Smith reached the prime minister's wife, but she would not disturb him.

Just as Smith was about to give up, the phone rang. It was the prime minister. It just so happened that the prime minister had a copy of Smith's book *Sea Fishes* and recognized Smith's name. Malan figured Smith must be calling about something important, so he called Smith back. Smith unloaded the whole story — of the first coelacanth, the lost innards, the long search for another specimen, and now the possibility of a second fish. It was, Smith urged, a matter of national prestige to get this fish.

Malan ordered an air force plane to take Smith to the Comoros. It was not the sort of mission the crew expected. The journey from Durban to Mozambique and then to little Dzaoudzi was a long jaunt for the DC-3 in those days. The trip only keyed up Smith further.

At last, the plane touched down. Smith first had to deal with formalities and meet the French governor of the Comoros Islands before he could see Hunt. Smith could barely contain himself and as soon as he could, he jumped into a car to race to the dock. Hunt had the fish in a large, coffin-shaped box. Smith was at first frozen with fear that Hunt had the wrong fish and that all of their trouble had been for nothing.

Then Smith opened the box. He knelt down to get a closer view. Tears were running down his face. He later recalled, "Fourteen of the best years of my life had gone in this search and it was true; it was really true. It had come at last."

Smith quickly regained his senses. He knew he had to get the coelacanth on the plane before the French authorities changed their minds about surrendering the fish. He also decided that it would be a good idea for the prime minister to be the first person to see the prize. After visiting the prime minister, Smith scooped up Marjorie and took the fish back to his laboratory.

There, Smith was delighted to dissect the beast. He found that the fish had no lungs and no air bladder as in other fishes, but had a swim bladder filled with fat. He also found evidence in the intestines that the coelacanth was a powerful predator. That and the fact that the fish was caught on a hand-held line in about twenty to thirty meters of water, just two hundred meters offshore, told Smith he was right about the coelacanth's lifestyle as a reef- and rock-dwelling predator — the fish was not a deep ocean species.

Many more coelacanths were caught in subsequent years, providing specimens for both scientists and museums. The fascination with this living fossil also inspired divers and ocean explorers to see it in its native habitat (Figure 7.5). Its swimming motion is particularly fascinating, as it moves its fins in alternating fashion — the right front fin with the left rear fin, the left front with the right rear — just as four-legged animals walk. Smith wrote a book about his adventure with coelacanths, dubbing the fish "Old Fourlegs."

With hours and hours of live observation, and many specimens caught over sixty years, one might think that scientists would have learned about all they could about coelacanths. Not quite.

Déjà vu?

In September 1997, Mark and Arnaz Erdmann were on their honeymoon, strolling through the seaside city of Manado on the northern tip of

FIGURE 7.5 *A coelacanth in its native habitat.*
Photo from JAGO submersible, Jurgen Schauer and Hans Fricke.

Sulawesi — one of Alfred Wallace's favorite islands in the Malay Archipelago (today part of Indonesia). (See Chapter 2, Figure 2.2 for map.)

Arnaz noticed it first. In a fish cart being pushed by an older man, there was a large, strange fish. She pointed it out to Mark and asked what it was. Mark, who had recently earned his Ph.D. in marine biology, immediately recognized it as a coelacanth.

There was no doubt in Mark's mind. He had read J. L. B. Smith's book about coelacanths when he was a young boy, and the fish had caught his imagination. But he wasn't sure if anyone had ever seen a coelacanth around Indonesia, some six thousand miles from the Comoros Islands.

Mark wasn't sure what to do. Was this a big discovery? Should he buy the fish and take some samples back to his home in California? He did have his sampling kit with him. He decided to take some pictures and look into Indonesian coelacanths later when he had more time.

"It was the biggest mistake I ever made," Mark later realized. He shared his photographs with several experts, each of whom concurred it was a coelacanth, and pointed out that the fish had never been found anywhere near those waters before. One expert suggested that perhaps a fishing boat might have off-loaded its catch from African waters in Sulawesi in order to avoid regulatory problems?

FIGURE 7.6 *The Indonesian coelacanth.*
Photo by Mark Erdmann, courtesy of Arnaz Mehta.

Mark consoled himself with the knowledge that he would soon return to Manado, where he could find the man with the fish cart and continue his investigation. But, once back in Manado, he had no luck at all finding the man or the fish. Thinking he may have missed a golden opportunity, he borrowed J. L. B. Smith's strategy and offered a reward and distributed posters of the fish to offshore fisherman. After some time passed and no coelacanth was turned in, Mark increased the reward. Then, one morning, almost a year after their honeymoon and the day in the fish market, Mark and Arnaz's boatman arrived, grinning.

Mark and Arnaz raced down to the beach where they found a fisherman holding a live coelacanth on a line in the shallow water. They guided the fish out into deeper water to revive it a bit. Arnaz jumped in with her diving gear and started swimming with the great fish. The fish was bleeding, and these waters had plenty of sharks, so the swim was a bit dangerous. But Mark and Arnaz had their proof (Figure 7.6). Coelacanths did live in Indonesian waters.

Two months later, *Nature* published Mark and Arnaz's discovery, and the coelacanth was again world news. It was not clear at first whether the Indonesian and African fish were the same species. They looked very similar in body form, but differed in color — the Indonesian form was brown with gold flecks.

Mark had one means of looking at coelacanths that J. L. B. Smith never had — DNA analysis. When the DNA of the Indonesian and African specimens were compared, they differed by 4.1 percent. This level of divergence is greater than that between chimpanzees and humans, and many other closely related species. The Indonesian coelacanth was a distinct species — Mark and Arnaz had found another living fossil.

Always Expect the Unexpected

Mark and Arnaz Erdmann's stroll through the Manado fish market and Marjorie Courtenay-Latimer's courtesy visit to the East London dock illustrate the role serendipity often plays in science. Marjorie recalled in an autobiographical note she penned some forty years later how several chance encounters led to her discovery:

> This story is one of the most astounding records of a woman's intuition, for:
> had I never gone to Bird Island;
> had I never met Dr. J. L. B. Smith who, of all the scientists I met as a young girl, struggling with meager funds in a small museum, always gave encouragement and never criticism;
> and had I not gone to the wharf to wish the men a Happy Christmas,
> there never would have been a coelacanth discovery in South Africa, on 22 December 1938.

It is true that discoveries are often quite accidental — occurring when least expected and turning up things that are not even remotely imagined. But, they would not be discoveries at all unless those who stumble upon something unexpected track down the meaning of what they have found.

CHAPTER QUESTIONS

1. "Living fossil" is a term that has been informally used to describe crocodiles, horseshoe crabs, ginkgo trees, and some other plants and animals. Are living members of these groups exceptionally long-lived species? How does the term "living fossil" apply to the coelacanth?

2. What is the significance of the structure of the coelacanth pectoral and pelvic fins?

3. What resemblances do you see between the fossil coelacanth and the fish sketched by Marjorie Courtenay-Latimer in Figure 7.2?

For more on this story, go to the *Into The Jungle* companion website at www.aw-bc.com/carroll.

PART IV

Evolution in Action

FIGURE 8.1 *Lesser flamingos at Lake Bogoria, Kenya.*
Photo by Rhonda Y. Kauffman.

A Sickle-Cell Safari

What may at first seem contrived or even ugly may be the best solution
that natural selection could devise.

—Francis Crick, *What Mad Pursuit* (1988)

Upper Gilgil, Kenya was a very long way from the scientific and educational hubs
of Europe and North America, but there could not have been a more perfect
place to inspire a young naturalist or anthropologist. Perched at 8,200 feet, over-
looking the Great Rift Valley, the surrounding landscape abounded with wildlife
and was home to a variety of tribes, each with their own language and traditions.

It was also a perfect spot to farm *Chrysanthemum*, whose flowers produce
the valuable natural insecticide *pyrethrum*. The plant, introduced into Kenya
by the British, thrives on the rich volcanic soil, ample rainfall, and abundant
sunshine in the highlands. It was on such a farm, at the edge of the forest, with
views of the massive volcano Mount Longonot and glittering Lake Naivasha,
that Tony Allison was raised.

Tony's father, in addition to establishing a successful farm, was well versed
in natural history and encouraged his son's budding interest. A family friend,

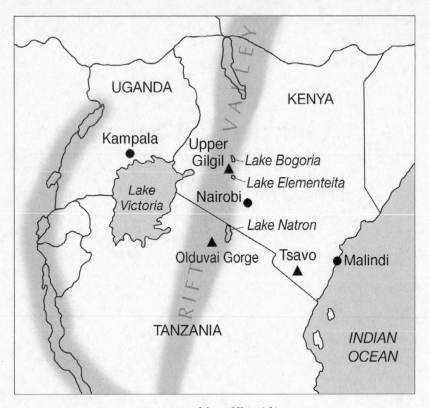

FIGURE 8.2 *Map of East Africa.*
The Rift Valley, prominent lakes and key locations are shown. *Drawn by*
Leanne Olds.

Leslie Brown, was the preeminent authority on African birds. On school holidays he took Tony on two-week long birding safaris. Leslie and Tony hiked down the fault scarp and backpacked to the shores of the soda lakes of Rift Valley, such as Lake Nakaru and Lake Bogoria (see Figures 8.1 and 8.2), which drew very large numbers of birds. Opening the tent flap in the early morning, they would be treated to the spectacle of thousands of pink lesser flamingos wading in the glassy, still water just fifty feet away. (Leslie later discovered the elusive breeding site of millions of East African lesser flamingos [consistency] at Lake Natron in Tanzania.) Tony worked as Leslie's field assistant, helping with observations and the preparation of birds to be sent to the Natural History Museum in London.

The archeological opportunities were just as rich as the bird watching. The great anthropologist Louis Leakey had established excavations close by, near Lake Elmenteita. Tony visited the dig site several times. Even though Tony was just a teenager at the time, Leakey was very polite and patiently explained his efforts to understand prehistoric Kenyan culture. Tony even got the chance to work on sorting out some of the stone tools and other artifacts of the so-called "Elmenteitan" culture that Leakey was then unearthing from a series of caves. Tony become fascinated with questions of human origins: Who were Kenya's earliest people? How were present-day tribes related to them and to each other? Tony dug into and was duly impressed by Darwin's *The Origin of Species* and *The Descent of Man*.

School vacations also meant trips to the Kenyan coast, which offered a dramatic change from the highland scenery. It was a two-day journey to Malindi. Tony and his family camped overnight at the Tsavo game park, under the watchful eye of its infamous lions, whose ancestors once had a taste for railway workers. The long, lovely beaches of Malindi offered great swimming, fishing, and surfing, but the coast also held some risks not present in the mountains. During one holiday, when he was about ten, Tony contracted malaria. His first experience with the disease left a long-lasting impression, as did the kindly doctor who treated Tony and prescribed some medicine. With the disease rampant across Kenya and other parts of Africa, medicine seemed like a noble calling. Tony's ambitions turned toward medical school.

After World War II broke out, Tony enrolled in college in South Africa, at the University of Witwatersrand in Johannesburg, to pursue his bachelor's and master's degrees in medical science. The school choice was partly Leakey's suggestion. The professor of anatomy at the university was Raymond Dart, who had discovered the fossil called the "Taung" child in 1924. Dart's interpretations of this fossil skull as that of a young bipedal hominid were very strongly contested because most scholars at the time thought that Asia, not Africa, was the cradle of mankind. Dart thought otherwise, but it would take twenty years or more for his view to be accepted and for *Australopithecus africanus* to take its place in the story of hominid evolution in Africa. Robert Broom, who discovered the second specimen of *Australopithecus*, lectured in Tony's anatomy course. His interest in the origins of humans as keen as ever, Tony took advantage of field expeditions to get some training in archeology and physical anthropology.

With heated debate accompanying the discovery of every bone, tooth, or stone, Tony thought that there must be better, more conclusive ways than conventional archeology and paleontology to get at the origins of humans and to work out the relationships between different peoples.

Out of Africa, and Back Again

In 1947, Tony enrolled at Oxford University to complete his medical training and soon became convinced that genetics offered a new way to get at human history. His teachers included some of the leading minds of population and evolutionary genetics. He was immersed in the mathematical works of R. A. Fisher, J. B. S. Haldane, and Sewall Wright, and impressed by Julian Huxley's account of the integration of genetics and evolutionary theory in *Evolution: The Modern Synthesis*. It was during his time at Oxford that Tony developed the notion that the use of genetic markers, such as human blood groups, would provide a better approach than linguistic or cultural traits to understanding human relationships.

His first opportunity to put his notion into practice came in the summer of 1949, between the completion of his basic science studies and the beginning of his formal medical training. Oxford University was mounting an expedition to Mount Kenya. While his colleagues would be studying insects and plants, Tony would pursue his anthropological interests by collecting blood samples from tribes all over Kenya. He hoped that blood types might reveal the genetic relationships among tribes. Before departing, he stopped in London for some advice on blood typing. An expert hematologist happened to mention that, in addition to the standard ABO, MNS, and Rh blood group tests, Tony should also test for the presence of sickle cells, as the frequency of that blood anomaly was reportedly unusually high in Africa. This casual suggestion would occupy Tony for the next decade.

Sickle-cell anemia was first described in 1910 by Chicago physician James Herrick who observed sickle-shaped red blood cells in the blood of an

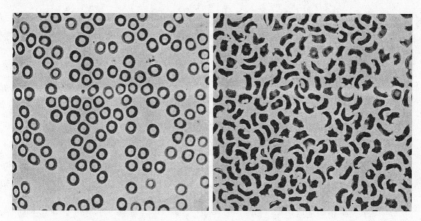

FIGURE 8.3 *A blood smear of an individual with the sickle-cell trait.*
This individual is heterozygous for the sickle-cell mutation. *Left*, under normal
oxygen concentrations, the red cells appear disc-shaped as normal. *Right*, but when
deoxygenated, the cells assume the sickle shape. *Photo from A. C. Allison (1956)*
Scientific American 195:87–94, *used by permission of author.*

anemic dental student. It was subsequently discovered that sickling of carriers' blood samples could be induced by allowing it to stand for a few days, or by treating the samples with the reducing agent sodium metabisulfate (Figure 8.3). By 1949, it was widely thought that carriers of sickle cell were heterozygous (carried one allele that coded for sickle cell and one allele that did not) and that individuals with sickle-cell anemia were homozygous (in whom both alleles coded for sickle cell). And in 1949, Linus Pauling's group at the California Institute of Technology demonstrated that hemoglobin from patients with sickle-cell anemia (Hemoglobin S, or HbS) had an altered charge, and thus sickle-cell was a "molecular disease." Heterozygotes generally had a mixture of wild-type hemoglobin A (HbA) and HbS, and their genotype was designated AS, while the genotype for homozygotes was designated SS.

Tony collected blood samples by visiting hospitals all over Kenya, from Lake Victoria to Mombasa and Nairobi. These samples included specimens from members of the Kikuyu, Luo, and Masai tribes. Overall, the blood group marker data from Tony's survey and those of other workers were both surprising and a bit disappointing. With the exception of the Masai, blood

TABLE 8.1 *Frequency of HbS in Selected Kenyan Tribes*

Tribe	Ethnic Affinity	District/Region	% HbS
Luo	Nilotic	Kisumu (Lake Victoria)	25.7
Suba	Bantu	Rusingo Island	27.7
Kikuyu	Bantu	Nairobi	0.4

group frequencies were fairly uniform among East African tribes and did not reveal informative differences.

However, Tony did notice great variation in the prevalence of the sickle-cell trait. In tribes living on the coast or near Lake Victoria, the frequency of sickle-cell carriers often exceeded 20 percent, but in tribes living in arid Central Kenya or in the highlands, it was usually less than 1 percent (Table 8.1). These associations held across regions with different languages and cultures, and were entirely independent of the blood group types he had documented.

The findings were puzzling for two reasons. First, because sickle-cell anemia is frequently lethal, how could it be that the frequency of the AS genotype was so high? And second, why would it be high in some areas and not others?

Tony had a flash of insight. Lake Victoria and the Mombasa coast were low-lying humid regions plagued by very high levels of malaria, which is carried by mosquitoes and caused by the parasite *Plasmodium falciparum*. Nairobi was at a higher elevation, had fewer mosquitoes, and thus did not have a high incidence of malaria. Because the parasite multiplies within red blood cells, perhaps the HbS allele, in heterozygotes, conferred some degree of resistance to malaria?

If true, Tony realized this would be the first ever demonstration of natural selection operating on a genetic variant in humans. Indeed, in 1949 there were no examples of natural selection in any species where a variant molecule was known. The link between malaria and the prevalence of the sickle-cell trait was a terrifically exciting idea, but Tony was in no position to test it. He had no time — his formal medical training was about to begin. He had no medical qualifications — he was still just a student. And he had no training — especially in parasitology. He decided not to publish the sickle-cell data or his theory. He would sit on his idea until he could earn the necessary credentials and training to test it. That would make for an agonizingly long wait of more than three years. He would have to live on tenterhooks, dreading that somebody else would realize the same connection and beat him to the punch.

There was one tremendous benefit from the idea burning in his head: It gave him a sense of urgency and enthusiasm for some subjects that his fellow students often lacked. He could not get enough of genetics, and the course in tropical medicine, one of the least popular among his classmates, was his favorite. The course was taught by Lt. General William McArthur, former director of the Indian Medical Service and an expert in tropical diseases. Tony clung to his professor's side just as he had to Leslie Brown on safari and to Louis Leakey at his excavation sites. Tony was always a willing assistant in McArthur's class, handing out microscope slides for his classmates to look at, and then staying after class for hours of one-on-one tutoring in tropical medicine. He loved every minute of it.

Finally, in 1952, he received his M.D. degree. He had not yet been scooped. It was finally time to put his training into practice.

There was just one not-so-tiny problem — Tony had no laboratory or position. Fortunately, in the course of his blood typing studies, he had met Alan Raper, the director of the Medical Laboratory in Kampala, Uganda. Raper was impressed with Tony's work and generously offered him the use of his house, his laboratory, and even his cook, while he was away on leave in 1953. It was perfect.

A Sickle-Cell Safari

Tony established Kampala as his base of operations and spent most of 1953 testing his idea that sickle-cell carriers (AS heterozygotes) were relatively resistant to malaria. He realized early on that he would have to take two factors into consideration that might complicate the situation: the induction of immunity by exposure to malaria and the widespread use of malarial drugs. In order to detect the role of the HbS mutation, he would have to try to minimize these factors in the design of his studies.

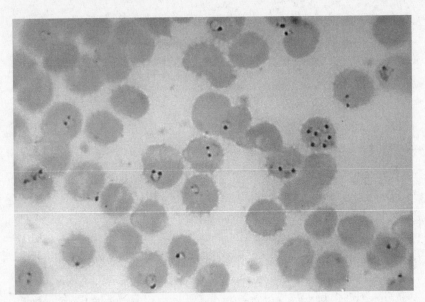

FIGURE 8.4 *A blood smear of an individual infected with malarial parasites.*
A red blood cell sample was treated with Giemsa stain to reveal parasites within cells
(black dots). *Photo courtesy of Dr. Mae Melvin, CDC Public Health Image Library.*

The first set of experiments he undertook was to see if AS heterozygotes
were relatively resistant to new infections. A pharmaceutical company had established a laboratory in Nairobi to test antimalarial drugs, and volunteers were rou-
tinely inoculated with *P. falciparum* as part of these clinical studies. In order to
try to minimize complicating factors, volunteers were selected that had not
recently visited malarial areas nor received antimalarial drugs.

Fifteen AS and fifteen AA Luo tribesmen were exposed to malaria and followed for forty days (then given a course of antimalarial drugs to arrest the
infection). Infection was established in fourteen out of fifteen AA individuals,
whereas only two out of fifteen AS individuals showed signs of infection in
blood smears (Figure 8.4). So far, so good.

Tony was concerned that these studies on adults were still affected by
prior exposures, so he decided to examine the levels of parasitemia in children. Consultation with various experts convinced him that children aged
six months to four years would be the most vulnerable to malaria infection
and the most informative for testing the hypothesis of the advantage of AS

FIGURE 8.5 *Masai woman and child on market day.*
Photo courtesy of A. C. Allison.

heterozygotes. In order to get an adequate amount of data for statistical purposes, he needed to examine a lot of children. Around Kampala malaria was at high levels and no preventative drugs were in use, so he could survey local children and use Raper's laboratory for the work.

On every weekday in the Buganda region around Kampala there were farmers' markets to which local women, accompanied by their small children, came to buy fresh produce (Figure 8.5). Every day Tony went out with a local pharmacist and offered a free check-up. He examined the children's eyes, chests, skin, and so forth and took a small sample of blood by pricking their finger or heel. Tony would then head back to the lab at 5:00 p.m. and spend until midnight preparing and analyzing samples, looking for HbS and parasite levels. After six months of exhausting work, he had data for 290 children, and the results he hoped for.

The incidence of parasitemia was 27.9 percent in the AS heterozygote children and 45.7 percent in the normal AA homozygote children. Furthermore, parasite density in the AS heterozygote children was lower than in the AA homozygote children. These results suggested that AS heterozygote children had a lower incidence of malaria or were affected for shorter periods than AA homozygote children, and thus AS heterozygote children would have a selective advantage in regions where malaria was at high levels.

Again, so far, so good. But while the Luo and Kampala studies supported Tony's theory that malaria and sickle-cell anemia were connected somehow, he wanted to know if the association between the high frequency of the HbS allele and high levels of malaria held *everywhere*.

Although Tony had data that showed the correlation from his 1949 Kenyan expedition, he felt he needed to test more tribes throughout East Africa. He embarked on what might be best described as a "sickle-cell safari." The journey took him from the forests of Western Uganda to the coasts and highlands of Tanganyika (now Tanzania) and Kenya, and he even managed to pay a visit to Louis Leakey's ongoing dig at Olduvai Gorge. Tony used the facilities of district hospitals where he could; otherwise his camp was his laboratory, with the crucial piece of equipment being a battery-operated microscope inside his tent (Figure 8.6). Tony managed to test five thousand East Africans in total, representing three countries and more than thirty different tribes. He found HbS allele frequencies of up to 40 percent in some areas where malaria was hyperendemic, and frequencies as low as 0 percent in areas where malaria was absent.

He had all the confirmation he needed.

Disease and Natural Selection on Humans

Large differences in sickle-cell allele frequencies among Kenyan and Ugandan tribes had been noted in surveys conducted by other researchers.

FIGURE 8.6 *On the sickle-cell safari.*
Tony Allison's camp under a fever-thorn tree (*Acacia xanthophloeia*), so-called
because they grew near the water in malarious regions. *Photo courtesy of A. C. Allison.*

However, these scientists attributed the differences to different factors. Some researchers thought that the differences reflected tribal migration histories, or the degree of genetic mixing among tribes, or a very high genetic mutation rate, or even a higher rate of reproduction among families of sickle-cell carriers to compensate for child mortality caused by the disease. Only Tony had looked for, and found, a natural environmental explanation, one that cut across tribal lines.

With all of the pieces of the malaria and sickle-cell trait connection fitting together, Tony, in a series of three articles published in 1954, made the case for malaria as an agent of natural selection on humans. In an article in the *British Medical Journal*, Tony explained:

> The proportion of individuals with sickle cells in any population, then, will be the result of a balance between two factors; the severity of malaria, which will tend to increase the frequency of the gene, and the rate of elimination of the sickle-cell genes in individuals dying of sickle-cell anaemia . . . genetically speaking, this is a balanced polymorphism, where the heterozygote has an advantage over either homozygote.

How much of an advantage was the sickle-cell gene to AS heterozygotes? This was very important to determine in order to understand the exceptionally high frequency of the sickle-cell trait in some regions. To estimate the selective advantage, he leaned on his Oxford training in population genetics, and the help of a collaborator and outstanding mathematician, Sheila Maynard Smith. Tony and Sheila reasoned that if they could figure out the *disadvantage* of SS homozygotes, then from basic population genetic principles they could estimate the advantage AS heterozygotes must enjoy that would account for their high frequency in malarial areas.

Tony's firsthand experience with and data on the incidence of the sickle-cell trait in children and adults was crucial. In the Luo tribe, Tony had measured a frequency of 25.7 percent for the sickle-cell trait. Assuming this was a mixed group of AS heterozygote and SS homozygote individuals, and that all SS homozygotes survived, Tony calculated that the expected ratio of SS homozygotes to all sicklers was about 1:12. However, he found that the actual ratio was closer to 1:35, meaning that only about one-third of SS homozygotes survived to young adulthood with the potential to have children. Other scientists estimated from other data that about 20 percent of SS homozygotes survived and reproduced. The reduced fitness of SS homozygotes meant that the overall frequency of the sickle-cell allele would be rapidly reduced, unless there was some advantage to the AS heterozygote. Sheila Maynard Smith calculated that advantage to be about 26 percent. In other words, in high malarial zones, 26 percent more AS heterozygote children than AA homozygote children reached adulthood. That is a whopping selective advantage, on par with some of the largest selective advantages ever measured for any trait in any species.

As word of Tony's discovery began to spread, leading evolutionary biologists wanted to hear more about it. Tony was invited to speak at the 1954 Cold Spring Harbor Symposium — the most influential meeting in the then just blossoming field of molecular biology (at the previous year's meeting James Watson and Francis Crick explained their model of DNA structure). It was Tony's first trip to America and a golden opportunity. Many of the current and future figures of population genetics and evolutionary biology were there,

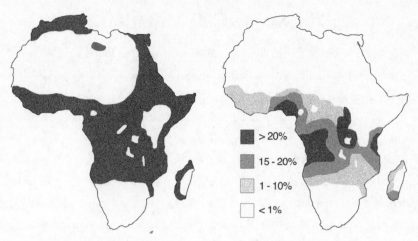

FIGURE 8.7 *The geography of sickle-cell hemoglobin and malaria.*
These maps show the close correspondence between the distribution of malaria (*left*)
and the frequency of the sickle-cell trait (*right*) across Africa. *Maps are based upon
A. C. Allison (2004) Genetics 66:1591. Redrawn by Leanne Olds.*

including Theodosius Dobzhansky, Ernst Mayr, Sewall Wright, James Crow, E. B. Ford, and Motoo Kimura.

Tony showed the audience how the frequency of the sickle-cell trait and the incidence of malaria coincided on a map of East Africa. In an otherwise math-heavy meeting, Tony's clear demonstration of the selective advantage of AS heterozygotes was a welcome change and very well received.

Such a large advantage, Tony went on to point out, also explained the high incidence of the sickle-cell trait outside of Africa. In the Lake Copais area of Greece, the trait reached a frequency of 17.7 percent, in part of India it reached nearly 30 percent, and it was also common in parts of Italy. Tony argued that these high frequencies and those in parts of Africa could not be due to ancestral relationships. Blood type markers did not support genetic relationships among these populations. So what did all of these populations have in common? They lived in regions notorious for a high incidence of malaria. That correlation could be mapped onto not just East Africa, but across all of Africa, southern Europe, and southern Asia (Figure 8.7). These maps would become textbook illustrations of natural selection on humans.

The Sickle-Cell Mutation

Tony speculated that the sickle-cell trait in different populations could result from each of those populations acquiring genetic mutations independently. But there was no way to test that idea at the time. The structure of DNA had only just been revealed, the genetic code was unknown, and methods for analyzing protein sequences were in their infancy.

Tony, and many others around the world, wanted to know the precise molecular defect in HbS. When Tony returned to England from his sickle-cell safari, he happened to meet Linus Pauling, whose team had demonstrated that HbS was different from HbA. Tony told Pauling about his not-yet-published discovery, and Pauling invited him to Caltech to work on the HbS protein.

There were few, if any, greater minds or greater personalities in twentieth century science than Linus Pauling. His seminal work on the nature of chemical bonds and his solution of the alpha-helical structures within proteins earned him his first Nobel Prize, in Chemistry, in 1954. His leadership in the opposition to atmospheric testing of nuclear weapons earned him his second Nobel Prize, for Peace, in 1962.

Tony's year in Pasadena was an eye-opening experience. He would often have breakfast at Pauling's house, which enjoyed a great view of the surrounding hills, then meet again later in Pauling's office to report the results from the day in the lab, and to have Pauling explain what they meant.

Tony soon decided that determining the sequence of the HbS protein was not yet feasible, so he focused instead on trying to figure out why HbS caused sickling. Pauling thought that HbS was aggregated in solution, while Max Perutz (a future Nobel laureate) thought that HbS actually crystallized inside red cells. Tony found out that Pauling was right, again, and that HbS aggregated into long rod-like structures.

Later, when back in England, Tony provided samples of the HbS protein to researchers who were using new techniques to determine protein sequences. In 1957, it was finally determined that HbS differed from HbA at just one amino acid, a valine in place of glutamic acid. Once the genetic code was deciphered and methods for the isolation and sequencing of DNA were developed, HbS was found to be due to a single base mutation (GAG → GTG) in the glutamic acid codon. Subsequently, the sickle-cell genes from different regions of Africa, southern Europe, and India revealed that this same mutation arose at least five different times, confirming Tony's hypothesis of the independent origins and spread of sickle cell in different malarial zones.

The Road from Gilgil

In the 1930s and early 1940s, paleontology, systematics, and population genetics became integrated into what was called the "Modern Synthesis" of evolutionary theory. One might think that many examples of natural selection on specific genes would have been known at the time, but that was not the case. Natural selection had been demonstrated in various ways since Darwin's original work, but there was nothing known about the identity or function of any genes involved. There was not a single "integrated" example of natural selection where the agent of natural selection was known, the effect on different genotypes could be measured, the genetic and molecular basis of variation was known, and the function of the gene or protein was understood. Who would have thought that with all of the talent and brainpower in evolutionary theory then populating the upper crust of academia, the credit for the first such example would go to a newly minted doctor and former farm boy from the Kenyan highlands? And that it would be demonstrated in humans, no less?

That first flash of insight during the 1949 Kenyan expedition carried Tony far, both geographically and professionally, in the ensuing six decades. After his stint in Pasadena, he spent twenty years directing a laboratory in England, then several years running a tropical disease laboratory in Nairobi, and then moved to California to join the pharmaceutical industry. There, he and his wife Elise Eugui played key roles in the development of what is today the leading drug used to prevent the rejection of transplanted organs.

Tony's discovery of the link between sickle cell and malaria resistance still stands out as a leading example of "evolution in action." This is because the agent of natural selection is known and still active, the degree of resistance is measurable and is clearly most important early in life, there is a simple genetic basis for resistance, and the association has been so well documented both geographically and clinically.

It is also a very important example because it challenges commonly held notions about mutation and evolution. Mutations are often perceived to be harmful, as is the HbS mutation under most conditions. However, as Tony put it, "the sickle cell mutation shows that mutation is not an unmixed bane to the human species . . . other mutant genes that are bad in one situation may prove beneficial in another." In fact, a few years after his HbS studies, Tony and David Clyde found that an enzyme deficiency caused by mutations in the gene encoding glucose-6-phosphate dehydrogenase (G6PD) affords considerable protection against *P. falciparum* malaria in Tanzanian children, a finding confirmed many times since. This explains why the enzyme deficiency is common in areas where malaria abounds.

Malaria has had a profound impact on human genetics and evolution that continues today. Over 40 percent of the world's population lives in malarial areas and over five hundred million cases occur each year, causing one to two million deaths. G6PD deficiency, HbS, and many other mutations now known to be associated with malarial resistance reveal that the battle against malaria has made a very strong mark on human evolution. Those mutations also demonstrate how natural selection works with whatever variation is available in the struggle for survival, and not necessarily by the best means imaginable.

We'll see in the next chapter how some species take even more radical measures in modifying their blood in the struggle to adapt to the challenges of their environment.

CHAPTER QUESTIONS

1. How did Tony Allison's early life experiences in Kenya prepare him to make the discovery of the sickle cell–malaria link?

2. What makes the sickle-cell mutation a balanced polymorphism?

3. Why was the demonstration of human resistance to malaria important to evolutionary biology?

For more on this story, go to the *Into The Jungle* companion website at www.aw-bc.com/carroll.

FIGURE 9.1 *The Norvegia at Bouvet Island.*
Photo from Fangst Og Forskning I Sydishavet *by Bjarne Aagaard, Volume 2,*
"Nye Tider." Published by Gyldendal Norsk Forlag, Oslo, 1930.

In Cold Blood:
The Tale of the Icefish

In all things of nature there is something of the marvelous.

—Aristotle

It was a long way just to go fishing.

The 125-foot converted wooden sealing boat *Norvegía* put to sea out of Sandeford Harbor, Norway on September 14, 1927. Its primary destination was perhaps the most remote piece of land on the planet. Tiny Bouvet Island, a speck in the vast Southern Ocean, lay more than six thousand miles from Norway, sixteen hundred miles from the tip of Africa, and more than three thousand miles from South America.

In the mid-1920s, commercial whaling was booming. The Norwegian invention of factory ships allowed greater numbers of animals to be taken without relying on shore facilities. Finding new stocks of whales was a priority for the entrepreneurs who went to sea, and establishing claims to territory and waters was a priority for the countries involved. The Norwegian government wanted to stake a claim to this ice-covered volcanic rock with

FIGURE 9.2 *Ditlef Rustad on the Norvegia foredeck.*
Photo from Fangst Og Forskning I Sydishavet *by Bjarne Aagaard, Volume 2,*
"Nye Tider." Published by Gyldendal Norsk Forlag, Oslo, 1930.

the aim of establishing some kind of outpost there that could assist the whaling fleet.

The *Norvegía* thus had a three-pronged mission — commercial, political, and scientific. It was outfitted and financed by a leading whaling businessman, but equipped largely for research. The young zoologist aboard, Ditlef Rustad, was to carry out investigations of marine life, paying close attention to whale populations (Figure 9.2).

After two months at sea, the *Norvegía* reached Cape Town, South Africa where it was prepared for the more treacherous leg of the voyage to Bouvet. The ship was equipped with a coal-fired steam engine, but could do just seven knots at best on a calm sea. The "Roaring Forties" and the "Furious Fifties," the names sailors had given to the wind- and storm-battered latitudes that lay further south, were anything but calm. The ship needed to take on all the coal it could in order to face the uncertain weather and to maximize its range as it explored Bouvet and its environs.

Sixty tons of coal were placed in sacks on deck, which soon created problems. The sacks made the ship top-heavy and exacerbated the rolling in large waves. When water poured onto the ship's deck in rough seas, the

FIGURE 9.3 *Claiming Bouvet Island for Norway.*
Taken on December 1, 1927. Ditlef Rustad is at the far left. *Photo from* Fangst Og
Forskning I Sydishavet *by Bjarne Aagaard, Volume 2, "Nye Tider." Published by*
Gyldendal Norsk Forlag, Oslo, 1930.

sacks prevented drainage, so many had to be thrown overboard. As the
Norvegía made its way south, the sea temperature dropped quickly and very
soon another hazard — icebergs — appeared. On November 30 the crew got
a first glimpse of Bouvet Island, but a snowstorm prevented the crew from
rowing ashore. Finally, on December 1, it was possible to land a party and to
erect a pole bearing the Norwegian flag (Figure 9.3). The first mission had
been accomplished.

Surveying of the waters around the island was constantly interrupted by
storms, rough seas, and hazards to navigation. On some occasions it was very
difficult for the shore boats just to return to the mother ship. On December 3,
during a storm, the *Norvegía* struck a rock off the coast of the island, which
tore off some of her ice protection, damaged the keel, and opened a leak.

Despite the hazards, the ship spent about a month around Bouvet. Rustad
spent as much time as he could trawling for plankton and fish with a net. On
the day after Christmas, at a depth of about one hundred feet, he caught some
unusual looking fish he called "crocodile-fish" on account of their large, pro-
truding jaws full of teeth. About twenty inches long, they had large pectoral

and tail fins, but were very pale, almost transparent. When he sliced one open, Rustad immediately noted their most peculiar feature of all — their blood was colorless. He took some photographs but did not keep the fish.

The leak in the *Norvegía* grew to the point where hand pumps had to be used to keep her afloat. As the pack ice increased, and the coal supply decreased, the decision was made to head to South Georgia Island for repairs. Rustad eventually made his way back home to Norway. The strange fish were well out of his mind, until two years later, when another zoology student brought them up.

A Bloodless Fish?

Johan Ruud also traveled to the Antarctic, in 1929, aboard the factory ship *Vikingén*. One day on deck, one of the old hands said to him, "Do you know there are fishes here that have no blood?" Ruud knew that was impossible because he had learned from textbooks that one trait all vertebrates share is red blood, the color of which is due to the presence of oxygen-binding hemoglobin protein in red blood cells.

Thinking that the veteran was just having fun with him, Ruud replied, "Oh yes? Please bring some back with you."

When the shipmate returned empty-handed, Ruud dismissed the whole conversation as just shipboard lore. But when he returned to Norway in 1930 and mentioned the tall tale to Rustad, he was shocked when Rustad told him, "I have seen such a fish" and produced his earlier photographs to prove it. The crocodile fish or "icefish," as the whalers also called them, were very real.

Ruud, who subsequently became a professor of marine biology at the University of Oslo and leader of the Norwegian Institute for Whale Research, heard nothing more about the bloodless fish for nearly twenty years. Then in 1948, one of his students returned from an Antarctic expedition with some icefish he had caught in the Straits of Magellan. The student noticed that the gills were also white, unlike the red gills of other fish. His curiosity reignited, Ruud asked other colleagues voyaging to the Antarctic to be on the lookout for, or better yet, to bring back more icefish (Figure 9.4).

FIGURE 9.4 *An icefish.*
Adult mackerel icefish, *Champsocephalus gunnari.*

A few more samples trickled back to Norway, but these were always stored
in preservatives that limited the kinds of information Ruud could extract. He
needed fresh fish, and the only way to get those was to go back to the Antarctic
himself. In 1953, nearly twenty-five years after his first journey and the first
rumors of the icefish, Ruud went to the Southern Ocean as the leader of an
international whale-marking expedition.

A few days before Christmas, Ruud landed at South Georgia Island. He was
anxious to get some icefish and find out once and for all what was different
about them. In the first two days on South Georgia Island, before he could
even set up his makeshift laboratory, three specimens were brought to him.
He drew their blood and noticed right away that it was almost transparent
(Figure 9.5). He popped the samples into a refrigerator for later study, figuring
that he would have all the fresh icefish whenever he wanted. When he cen-
trifuged the blood, he obtained a pellet of cells comprising less than 1 percent
of the total blood volume. The plasma above was as "clear as water." Moreover, he could not detect a single red blood cell under the microscope. The
icefish, unlike all other vertebrates, completely lacked the pigmented oxygen-
carrying cells that had, until then, been found in every living vertebrate.

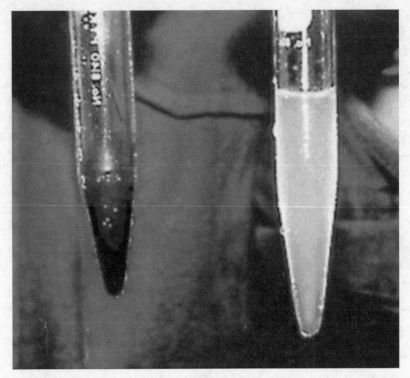

FIGURE 9.5 *Icefish blood.*
Compared with the red blood of a rock cod (left), icefish blood is white (right)
and contains only a small volume of white cells and no red cells. *Photograph from
B. D. Sidell and K. M. O'Brien,* Journal of Experimental Biology 209:1791–1802
(2006), used by permission.

Ruud was eager to measure the oxygen-carrying capacity of fresh icefish
blood and to compare it with red-blooded fish in the same waters. But, as luck
would have it, no more icefish beyond the first three appeared in his traps or
were hooked. Day after day passed as Ruud's scheduled departure date
approached. Desperate, he sent out an appeal to all of the whaling stations on
the island. To his delight and relief, a doctor at one of the two stations showed
up in the nick of time with a live fish in a barrel of sea water.

Ruud went right to work and drew blood from the heart. He took eight separate measurements and found that icefish blood had an oxygen capacity of
0.77 percent by volume, compared with capacities of around 6.0 percent for its

red-blooded relatives. The icefish blood contained no more oxygen than would be carried by liquid plasma alone.

Ruud had his answers and reported his findings in *Nature* in 1954. Even more than fifty years later, his discovery is a shock for any biologist reading it for the first time. The icefish, a group of sixteen species, are the only "blood-less" vertebrates to have ever been discovered.

As Ruud noted in his paper, "The fact that a good-sized vertebrate can exist without any oxygen-binding blood pigment raises some interesting questions." These questions included where, when, why, and how did these fish evolve? What happened to their hemoglobin? And, how could these fish survive without hemoglobin or red blood cells?

Ruud surmised that the answers to these questions would be linked to the icefishes' unique ecological setting: the very cold and stable temperature of the food- and oxygen-rich waters of the Antarctic. He believed that only in such a unique combination of circumstances "is the survival of these peculiar animals possible."

Answers to the mysteries posed by the icefish would not unfold for decades, until a new generation of scientists traveled to the Antarctic and brought new approaches to studying icefish history — by analyzing not just their blood, but their DNA. For the DNA of icefish and their relatives harbors a detailed record of how these remarkable animals have evolved — of the genetic changes that have shaped their unique history. These genetic changes offer some remarkable glimpses into "evolution in action."

A Matter of Degrees

Before confronting the mystery of the anemic icefish, biologists first confronted the mystery of why (and how) there are any fish at all in Antarctic waters. It came as a big surprise to some of the first explorers of the region that, contrary to the inhospitable habitat they imagined, the coldest waters in the world are alive with fish.

How cold are the Antarctic waters? The waters of McMurdo Sound have been found to average about –1.9°C (28.6°F), and to vary by only about 0.1°C with depth or through the seasons. Most tropical and temperate fishes freeze at around –o.8°C, so Antarctic fish must have evolved some means to cope with the icy Southern Ocean.

One group of fish, the suborder *Notothenioidei*, of which icefish constitute one family, dominates Antarctic waters. Notothenioids comprise about one-third of all Antarctic fish species and about 90 percent of Antarctic fish biomass.

Notothenioids are unknown from the Antarctic fossil record as recently as forty million years ago, when the Antarctic coastal waters abounded with sharks, rays, catfish, and other groups of fish that are now vanished. This dramatic turnover in fish fauna is associated with dramatic changes on land and in the ocean. About thirty-three to thirty-four million years ago, Antarctica became fully separated from South America and completely surrounded by deep ocean. Ensuing changes in ocean currents formed what is called the "Antarctic Convergence," which isolated the waters around the Antarctic and limited the inflow of warm waters and the migration of fish from northern waters (Figure 9.6). It is thought that over the span of twenty million years or so, the water temperatures around Antarctica plummeted from roughly 10°C (50°F) and reached below freezing by about fifteen million years ago. The notothenioids have thrived in this icy habitat, which leads to the question, "How do these fish endure the extreme cold and avoid freezing?"

The biologist who has made the most contributions to our understanding of the adaptation of notothenioids to the cold Southern Ocean is Arthur DeVries. Now a Professor at the University of Illinois, DeVries has ventured to the Antarctic more than forty times since 1961, beginning when he was a graduate student. DeVries grew up on a Montana farm which, while it did not prepare him for long sea voyages, gave him plenty of experience of working in the cold and snow.

From Ross Island, a small volcanic island forty miles from mainland Antarctica, DeVries and his colleagues launched a series of studies into the temperature tolerance of Antarctic fish. They caught the fish in traps and with conventional hooks and lines in holes cut through fifteen feet of sea ice. They had to make their holes over deep water to avoid the Weddell seals that

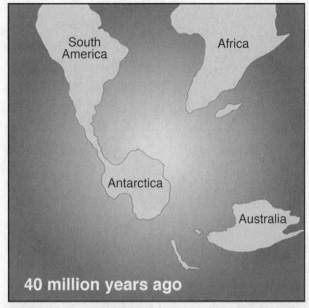

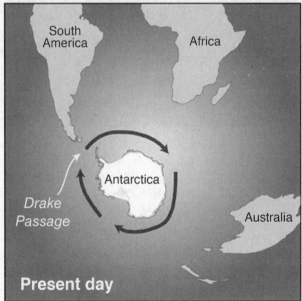

FIGURE 9.6 *The changing geology of Antarctica.*
When Antarctica became fully separated from South America, this opened Drake's
passage and changed ocean currents such that Antarctic waters became very cold
and isolated. *Illustration by Jack Cook, Woods Hole Oceanographic Institution.*
Redrawn by Leanne Olds.

FIGURE 9.7 *Fishing in the Antarctic.*
Top, Art DeVries standing on the ice sheet over McMurdo Sound, Antarctic. The
fishing hut is behind him, and Mt. Erebus is in the background. *Bottom*, Art DeVries
(checkered jacket), fishing through a large hole for antifreeze-bearing fish.
Photos circa 1961, courtesy of Art DeVries, University of Illinois.

ruined holes made in shallower water (Figure 9.7). Several species of
notothenioids, some reaching more than fifty pounds, were landed.

DeVries found that the sera of several notothenioids froze at –2.0 to 2.1°C,
just a fraction of a degree below the freezing temperature of the sea water, and
well below the freezing point (–0.8°C) of temperate fish. This extra degree or
so of freezing-point depression means that notothenioids do not freeze at water
temperatures that would freeze other fish.

DeVries and his colleagues then analyzed notothenioid sera in order to identify the components responsible for the freezing resistance. They found that a glycoprotein fraction of the serum was responsible for about 50 percent of the freezing point depression, with dissolved salts (NaCl, urea, and other compounds) responsible for the remainder. The notothenioids are chock full of these antifreeze glycoproteins (AFGPs), containing twenty to thirty-five milligrams of antifreeze protein per milliliter of serum (about half of all serum protein).

Subsequent structural studies revealed that AFGPs have a very unusual and simple structure. They are composed of four to fifty-five repeats of just three amino acids, threonine-alanine-alanine (sometimes threonine-proline-alanine), with a disaccharide (sugar) of N-acetylglucosamine and galactose attached to the threonine residue.

The main enemy of fish in polar waters is not so much the cold, but ice. In the $-1.9°C$ water column there are small crystals of ice that, if they enter the fish via the gills or by ingestion, can nucleate the formation of larger ice crystals and freeze the fish. The AFGPs work by adsorbing to small ice crystals and lowering the temperature at which crystals can grow.

Warm-water fish do not have AFGPs, so the antifreeze genes must have been somehow invented by notothenioids. One of the major quests in evolutionary biology has been to understand how new gene functions arise, and the AFGPs presented a perfect opportunity to explore this question. It took more than twenty-five years after the discovery of AFGPs to find out, but it was well worth the wait.

When a team at the University of Illinois, including Chi-Hing Cheng, Arthur DeVries, and Liangbiao Chen, isolated the AFGP genes from a giant Antarctic toothfish (*Dissostichus mawsoni*), they noticed that parts of one AFGP gene had striking similarities to parts of another gene that encodes trypsinogen, a digestive enzyme. The first exon of the AFGP, including the 5′ untranslated sequence and signal peptide, was 94 percent identical in sequence to those of the trypsinogen gene, the 3′ end of the AFGP gene was 96 percent identical to trypsinogen, and two introns were 93 percent identical. These are extraordinary degrees of identity among two functionally distinct genes and suggest that the AFGP gene evolved from an ancestral trypsinogen

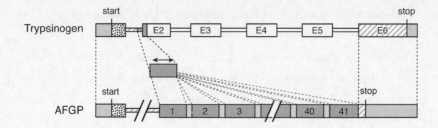

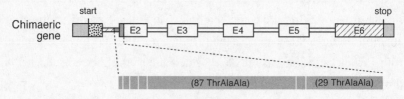

FIGURE 9.8 *The origin of the antifreeze genes.*
The AFGP genes evolved from an ancestral trypsinogen-like gene. *Top*, the structure of a notothenioid trypsinogen-like gene. The first exon is shaded and stippled, the second through fifth exons (E2–E5) are indicated, and the start and stop codons are labeled. *Middle*, the structure of an AFGP gene. The first and last exons bear very strong similarities to the first and sixth exons of the trypsinogen-like gene. The AFGP repeats in exons 1–41 are encoded by a nine base repeat that is homologous to a sequence spanning junction between the first intron and second exon of the trypsinogen gene. These sequence similarities indicate this AFGP gene was derived from a trypsinogen gene. *Bottom*, striking evidence of the origin of AFGP genes from the trypsinogen-like ancestral gene comes in the form of a chimeric gene that contains all of the trypsinogen exons together with AFGP repeats inserted at the junction of intron 1 and exon 2. *Figure is modified from C.-H. Cheng and L. Chen (1999)* Nature 401: 443–444. *Drawn by Leanne Olds.*

gene. Furthermore, the team noted a nine base pair segment straddling a splice junction in the trypsinogen gene that encoded a Thr-Ala-Ala tripeptide, the repeat building block of AFGPs. The team surmised that the many tripeptide repeats of the AFGP gene evolved by repeated duplication of DNA containing this short segment.

This model for the origin of AFGP received the best confirmation imaginable when Cheng, DeVries, and Chen later isolated another AFGP gene that was clearly a chimeric gene that encoded *both* AFGPs and trypsinogen (Figure 9.8). They had captured an evolutionary intermediate between an ancestral trypsinogen gene and an evolving AFGP gene. This discovery provided a very rare view of how a preexisting gene gave birth to a new gene and a new protein with a new function — one that ultimately allowed an entire group of fish to invade and dominate perhaps the most challenging habitat on Earth.

The origin of the antifreeze proteins stands out as a prime example of how evolution works more often by "tinkering" with materials that are available — in this case the sequence of an already existing gene — rather than by inventing new functions completely from scratch. It is also a stellar illustration of how species' DNA contains a record of their evolutionary history. Examination of the icefish DNA record would also deliver some big surprises.

Fossil Genes

The riddle of what happened to icefish hemoglobin also had to wait a long time before it was solved. More than forty years after Ruud first analyzed and described icefish blood, teams led by two Antarctic veterans, Bill Detrich of Northeastern University and Guido di Prisco of the Consiglio Nazionale delle Ricerche (Naples, Italy), probed icefish DNA for their adult α-globin and β-globin genes. In red-blooded fish, these two genes are located next to one another in a head-to-head orientation. But in all icefish species examined, only a remnant of the α-globin gene was detected and no β-globin sequences were detectable (Figure 9.9). The β-globin gene and all but the third exon and a bit of the second intron of the α-globin gene were gone — extinct. The remaining chunk of the α-globin gene is a molecular fossil, or pseudogene, a vestige of the red-blooded way of life in icefish ancestors that was abandoned.

Red blood cells and globin genes are found in all other fish and vertebrates, which tells us that hemoglobin has nurtured vertebrate respiration for over five hundred million years. How and why can icefish do without it?

As Ruud surmised, the answers to these questions pivot on the cold, oxygen-rich Antarctic waters. The frigid cold presents a variety of challenges to body physiology. At low temperatures, one difficulty is the increased viscosity of body fluids. Elimination of red blood cells certainly reduces the viscosity of blood and makes it easier to pump. But there remains the problem of

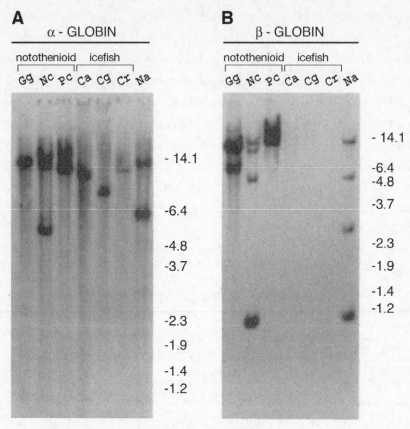

FIGURE 9.9 *The loss of globin genes in icefish.*
Southern blots of genomic DNA from three red-blooded notothenioid species (Gg,
Nc, Pc) and three icefish species (Ca, Cg, Cr) were probed with cDNAs for α-globin
(A) or β-globin (B). Size standards are on the right of each blot (Na). Note the
absence of any signal with the β-globin probe in the icefish. *Figure modified from*
E. Cocca et al., Proceedings of the National Academy of Sciences USA, 92:1817–1821
(1995); *copyright 1995 the National Academy of Sciences, USA.*

oxygen delivery — how do the icefish get enough oxygen to their tissues?
Several features of the icefish cardiovascular system seem to meet this challenge. Icefish have relatively large gills and have evolved a scaleless skin with
unusually large capillaries. These features increase the absorption of oxygen
from the environment. Icefish also have much larger hearts and about four
times larger blood volumes than fish of comparable size. It appears that icefish
compensate for the lack of active oxygen transport by circulating their dilute

hemoglobinless blood more often than their red-blooded cousins. Icefish hearts differ in another obvious and peculiar way — they are often pale. The reddish color of vertebrate hearts (and muscles) is due to the presence of myoglobin, a protein which binds oxygen more tightly than hemoglobin and sequesters it in muscles so that it is available upon exertion. But unlike most other fish and vertebrates in general, myoglobin is absent from icefish muscles and the hearts of several, but not all, icefish species.

Another Antarctic veteran, Bruce Sidell, from the University of Maine, and his colleagues have found that the myoglobin genes of pale-hearted icefish also bear crippling mutations. Curiously, different mutations have been identified in different species: in two species there is a five base pair insertion that disrupts the reading frame of the gene; in another species there is a fifteen base pair insertion upstream of the gene; and in a yet another species there is a mutation in the mRNA polyadenylation signal sequence downstream of the stop codon. In these species, the myoglobin gene is a molecular fossil.

Because the three myoglobin gene mutations are different, it appears that myoglobin function or expression has been lost at least three separate times within the icefish family. The presence of myoglobin in the hearts of ten species reveals that the function of myoglobin in icefish is probably in the process of being lost. One explanation for the incomplete loss of myoglobin in icefish may be the relatively young age of the icefish family (estimates range from two to eight million years). Even if myoglobin is not essential to icefish physiology, there may not yet have been enough time for mutations to inactivate the gene in all icefish lineages.

Whatever the fate of the myoglobin gene in the remaining species of icefish, it is clear that several species of icefish are getting along just fine without two ancient oxygen-carrying proteins that have nurtured vertebrate life for over half a billion years.

The DNA Record of Evolution

Paleontologist George Gaylord Simpson once stated that "species evolve exactly as if they were adapting as best they could to a changing world, and not at all as if they were moving toward a set goal." The evolution of the notothenioids and of the icefish provides vivid evidence for this axiom and how the earth and life evolve together.

These fish evolved from warm-water, red-blooded ancestors ill-suited to life in frigid water. But neither the changes in the Southern Ocean nor the adaptation of fish to them were abrupt. The evolution of notothenioids and icefish involved many genetic steps in addition to the invention of new genes and proteins and the destruction of some old genes seen here. Much of this history is also recorded in their genomes. Adaptation to cold temperatures places many demands on cell and organ physiology, and evidence of functional evolutionary changes in many other icefish and notothenioid proteins have been revealed. These include, for example, modifications of the thermal stability of the tubulin proteins involved in microtubule assembly.

By comparing the genes of different icefish, red-blooded notothenioids, and other fish, we can deduce when certain changes occurred during different intervals of icefish history, and place these in the context of geologic history. All two hundred or so Antarctic notothenioids have AFGPs and cold-stable tubulins, so those were early inventions in a common ancestor of all notothenioids that originated approximately fifteen to twenty-five million years ago. (Thick sea ice began to form about fourteen million years ago.) But only the icefish have fossilized globin genes, which means that the mutation(s) that inactivated those genes occurred in a common ancestor of all of the icefish that originated approximately two to eight million years ago. The changes in myoglobin genes are still more recent than the origin of icefish. The DNA record reveals that the icefish and its ancestors evolved in many steps.

There is a very important corollary to Simpson's axiom: What is good under one set of circumstances may not be so under other circumstances. The DNA record of the icefish, as well as that of other species, documents the somewhat messy course of evolution at the DNA level. The icefish fossil globin and myoglobin genes shatter any preconception that evolution is necessarily a "progressive" process. In the course of evolution, some capabilities and information are also lost. As lifestyles evolve, organisms' reliance upon different physiological systems and protein functions also shift, and former ways of living and the genes necessary for those ways of life can be abandoned entirely.

The DNA record reveals how natural selection acts only in the moment. It cannot preserve what is no longer used, and it cannot plan for the future. The icefish have made a remarkable evolutionary journey, but with respect to oxygen transport, they have bought a one-way ticket. They are ill-suited to life in warmer waters. Just as Johan Ruud deduced as to their origin, their fate, too, is tied to their unique, cold, oxygen-rich habitat.

CHAPTER QUESTIONS

1. How did fish antifreeze originate?

2. What is a fossil gene?

3. How do icefish obtain their oxygen? What might happen to icefish if the waters around the Antarctic became warmer?

4. It is sometimes said that Earth and life evolve together. How does the history of the icefish support this statement?

For more on this story, go to the *Into The Jungle* companion website at www.aw-bc.com/carroll.

General Review and Discussion

The stories here were chosen from among many great achievements in evolutionary science. While the time, place, and substance of the discoveries described differ a great deal, they do have some common themes with respect to how scientific exploration is conducted, how discoveries are made, and how new ideas are born and grow. Here are a few more questions and points to consider regarding themes that cut across many or all of the chapters.

1. What personality traits and experiences do you think are common ingredients in becoming a naturalist? A scientist?

2. What role does serendipity play in scientific discovery? Which stories and scientists here illustrate that role?

3. Darwin referred to *The Origin of Species* as "one long argument." In the stories here, consider how often scientific progress was a matter of discovering something new, and how often scientific progress was a matter of persuading others of its reality or significance.

4. Despite one hundred fifty years of scientific study, it is still claimed by many that "there is no evidence" for evolution. What elements of the stories here would you use to refute that claim?

5. George Gaylord Simpson said, "Species evolve exactly as if they were adapting as best they could to a changing world, and not at all as if they were moving toward a set goal." Which stories and evidence here support that statement?

Acknowledgments

A long time ago, I sat in my first college biology class. It was also my first day at college so I did not know what to expect. The professor strode in and I grabbed my notebook in anticipation of a blizzard of information. Instead, that professor, Dr. David L. Kirk at Washington University in St. Louis, just smiled, opened a copy of Darwin's *Voyage of the Beagle*, and began to read aloud to the class. My classmates and I soon relaxed, put down our pens, and just listened.

I learned a lot more from Dr. Kirk that day than where Darwin sailed or what he saw. First, as I hung on every word, I learned that I was in the right place, and that I definitely wanted to become a biologist. For that timely injection of inspiration, I am forever grateful. Second, my ability to recall that one class for all of these years has taught me that storytelling makes events memorable. That lesson has led to me telling stories in my classes, then writing them in books, and the writing of this book in particular.

Those sure were big payoffs for one hour, and maybe there is even a third lesson in my tale: Don't cut class! You might miss the most important stuff! And by the way, as I recall, none of that first day's material appeared on an exam.

The stories here would not have been told without the generosity of many people. My wife, Jamie Carroll, offered great encouragement at the inception of this undertaking, and keen critical input through to its completion. Jamie read every chapter from its earliest drafts onward, provided invaluable guidance on making the stories interesting and readable, and contributed art to several chapters.

I am especially indebted to Tony Allison for sharing his experiences in interviews and letters, and for providing photographs and other documents relating to his journeys in Africa. It has been a privilege to learn about his extraordinary life and accomplishments. Special thanks also to Art DeVries for generously sharing photographs and documents of his amazing and still ongoing work in the Antarctic.

I am also very grateful for the generous creative, critical, and logistical contributions from colleagues at the University of Wisconsin-Madison. Leanne Olds created most of the original artwork for the book and designed the cover; Steve Paddock provided detailed comments and suggestions on the entire manuscript; and Megan McGlone secured the permissions for the images obtained from other sources. I was also particularly fortunate to have liberal access to the holdings of the University of Wisconsin library system. The researching of this book led me to many old and rare volumes; many thanks to Elsa Althen for providing access to the rare books collection of the Biology library.

I also thank my colleagues at Benjamin Cummings/Pearson Education for their support and efforts in publishing the book. I am very grateful to Beth Wilbur for her commitment to and vision for this endeavor, to my editor Susan Teahan for steering me through the development process, and to Camille Herrera for speeding the book through production. Special thanks to the cadre of reviewers who provided very thoughtful and helpful comments on draft chapters:

Robert Dennison, Houston ISD
Nancy Monson, West Linn High School
Nancy E. Ramos, Northside Health Careers High School
Lori Nicholas, New York University
Joel Stafstrom, Northern Illinois University
Kathy Gallucci, Elon University
Roxanne Fisher, Chatham University
Gary Wellborn, University of Oklahoma
Anna Bess Sorin, University of Memphis
Patrick Enderle, East Carolina University
Cara Shillington, Eastern Michigan University
Craig Jordan, University of Texas at San Antonio
Michele Shuster, New Mexico State University
Lois Borek, Georgia State University
Barry Condron, University of Virginia
Barbara R. Beitch, Ph.D., Quinnipiac University

References and
Further Reading

The many sources used for these stories, including books, articles, websites, and interviews, are listed below. The source of each quote is identified by page number. Some references that are recommended for further reading on particular topics are indicated with an asterisk.

Chapter 1
Reverend Darwin's Detour

Thanks to the efforts of many scholars, and Darwin's habit of keeping his correspondence, there is a massive amount of information available on Darwin's life and work in the form of his diaries, letters, notebooks, and writings. Much of this material can now be accessed online at http://darwin-online.org.uk/. There are also many fine biographies of Darwin, two of which I drew upon here are:

*Browne, Janet. (1995). *Charles Darwin: Voyaging*. New York: Alfred A. Knopf.
*Desmond, Adrian, and James Moore. (1991). *Darwin: The Life of a Tormented Evolutionist*. London: Michael Joseph.

In the following list, the sources of various quotes or information are:

Barlow, Emma Nora. (1963). "Darwin's Ornithological Notes." *Bulletin of the British Museum (Natural History) Historical Series*, 2:201–273. (DON)
Barlow, Emma Nora. (1967). *Darwin and Henslow: The Growth of an Idea. Letters 1831–1860*. Berkeley and Los Angeles: University of California Press. (DH)
Burkhardt, F. H., and S. Smith et al., eds. (1985). *The Correspondence of Charles Darwin*. Cambridge: Cambridge University Press, UK. (CCD1)

Darwin, Charles. (1839). *Journal of Researchers into the Geology and Natural History of the Various Countries Visited by H.M.S. Beagle Under the Command of Captain Fitzroy, R.N. from 1832 to 1836*. London: Henry Colbourn. (VB)

Darwin, Charles. *Notebooks "B," "D," and "E"*. Images online at http://darwin-online.org.uk. (NB, ND, NE, respectively)

Darwin, Erasmus. (1803). *Zoonomia: Or, the Laws of Organic Life*. Vol. 1. Boston: Thomas and Andrews.

Darwin, Francis, ed. (1887). *The Life and Letters of Charles Darwin, Including an Autobiographical Chapter*. 3 vols. London: John Murray. (LL)

Darwin, Francis, ed. (1909). *The Foundations of the Origin of Species. Two Essays Written in 1842 and 1844*. Cambridge: Cambridge University Press, UK. (E42 and E44)

Keynes, Richard D., ed. (1988). *Charles Darwin's Beagle Diary*. Cambridge: Cambridge University Press, UK. (BD)

Lyell, Charles. (1832). *The Principles of Geology*. Vol. 2. London: John Murray.

SOURCES OF QUOTES

1. p.6 "You care for nothing...," LL, autobiography, p. 32.
2. p.9 "Quite the most perfect man...," CCD1, p. 110.
3. p.10 "I think you are the very man they are in search of," DH, p. 30.
4. p.10 "...if you think differently from me...," CCD1, p. 132.
5. p.10 "Disreputable to my character...," CCD1, p. 133.
6. p.10 "...all the assistance in my power," CCD1, p. 135.
7. p.13 "...stewed... in warm melted butter," DB, p. 35.
8. p.14 "I formerly admired Humboldt...," DH, p. 55.
9. p.14 "We breakfast at eight o'clock...," CCD1, p. 248.
10. p.14 "...by far the most savage...," DB, p. 99.
11. p.15 "...taste and look like a duck," DB, p. 105.
12. p.15 "...cargoes of apparent rubbish," see DB p. 106 (Fitzroy narrative 2:106–107).
13. p.16 "I have scarcely for an hour...," DB, p. 131.
14. p.18 "May providence keep...," DB, p. 132.
15. p.18 "...my messmate, who so willingly...," DB, p. 140 (footnote Fitzroy narrative 2:216–217).
16. p.19 "I know not, how I shall be able to endure it," DH, p. 63.
17. p.19 "Turned out to be most interesting," DH, pp. 77–79.

18.   p.21 "I saw the spot where a cluster of trees...," VB, pp. 406–407.

19.   p.23 "the stunted trees draw little signs of life...," DB, p. 352.

20.   p.23 "paradise for the whole family of reptiles...," DB, p. 353.

21.   p.23 "two very large tortoises...," DB, p. 354.

22.   p.24 "pointing out to me as a youngster...," LL1, p. 225.

23.   p.26 "I loathe, I abhor the sea," CCD1, p. 503.

24.   p.26 "I have specimens from four...," DON, p. 262.

25.   p.28 "...it never occurred to me...," VB, p. 475.

26.   p.29 "...one is urged to look to common parent," DON, p. 277.

27.   pp.29 and 30 all quotes from notebook B from text at http://darwin-online.org.uk/

28.   p.30 "It is a beautiful part of my theory...," notebook E, p. 71.

29.   p.32 "...the greatest success my humble work...," CCD2, pp. 218–222.

30.   p.32 "That the author of those passages...," CCD2 (letter 545).

31.   p.33 "I have just finished my sketch of my species theory," E42, p. xxvi.

32.   p.33 "...dear Old Philosopher," LL1, p. 221.

Chapter 2
Drawing the Line between Monkeys and Kangaroos

BOOKS

Beddall, B. G. (1969). *Wallace and Bates in the Tropics: An Introduction to the Theory of Natural Selection*. London: Macmillan Co.

*Quammen, D. (1996). *The Song of the Dodo: Island Biogeography in an Age of Extinction*. New York: Scribner.

van Oosterzee, P. (1997). *Where Worlds Collide: The Wallace Line*. Ithaca, NY: Cornell University Press.

Wallace, A. R. (1890). *The Malay Archipelago*. London: Macmillan and Co.

*Wallace, A. R. (1905). *My Life*. New York: Dodd, Mead, and Co.

ARTICLES

Forbes, H. O. (1914). "Obituary: Alfred Russel Wallace, O. M." *The Geographical Journal* 43: 88–92.

McKinney, H.L. (1969). "Wallace's Earliest Observations on Evolution." *Isis* 60:370–373.

Wallace, A. R. (1855). "On the Law Which has Regulated the Introduction of New Species." *Annals and Magazine of Natural History* 16:184–196.

Wallace, A. R. (1857). "On the Natural History of the Aru Islands." *Annals and Magazine of Natural History* 20:473–485.

Wallace, A. R. (1858). "On the Tendency of Varieties to Depart Indefinitely from the Original Type." *Proceedings of the Linnean Society of London* 3:53–62.

SOURCES OF QUOTES

1. p.37 "I'm afraid the ship's on fire," *My Life*, pp. 303–312.
2. p.37 "It was now. When the danger...," *ibid.*
3. p.40 "Here and there, too, were tiger pits...," *The Malay Archipelago*, p.18.
4. p.40 "...it was rather nervous work hunting for insects...," pp. 18–19.
5. p.40 "...as there were many bad people about...," in A. R. Wallace, "Letter from Macassar, Celebes," *Zoologist* 15 (1856):5559–5560.
6. p.41 "Nature seems to have taken every precaution...," *My Life*, p. 394.
7. p.41 "I had seen sitting on a leaf...," *The Malay Archipelago*, pp. 257–258.
8. p.42 "Every species has come into existence...," "On the Law..."
9. p.43 "...that the present geographical distribution of life...," *ibid.*
10. p.43 "...which contain little groups of plants and animals...," *ibid.*
11. p.43 "...they must have been first peopled...," *ibid.*
12. p.43 "They could not be as they are...," *ibid.*
13. p.44 "...crossing over to Lombok...," *The Malay Archipelago*, p. 155.
14. p.46 "Let us now examine," "On the Natural History."
15. p.46 "...we can scarcely find a stronger contrast...," *ibid.*
16. p.47 "...some other law has regulated the distribution of existing species," *ibid.*
17. p.47 "...to think over subjects particularly interesting to me," *My Life*, p. 361.
18. p.48 "The life of wild animals is a struggle for existence...," "On the Tendency."
19. p.49 "I am a little proud," *My Life*, p. 366.

Chapter 3
Life Imitates Life

BOOKS

*Bates, H. W. (1892). *The Naturalist on the River Amazons, with a Memoir of the Author by Edward Clodd*. London: John Murray. (NORA)

The Correspondence of Charles Darwin. Vol. 9, 1861. Cambridge: Cambridge University Press, UK.

The Correspondence of Charles Darwin. Vol. 10, 1862. Cambridge: Cambridge University Press, UK.

The Correspondence of Charles Darwin. Vol. 11, 1862. Cambridge: Cambridge University Press, UK.

ARTICLES

Allen, G. (1862). "Bates of the Amazons." *Fortnightly Review* 58: 798–809.

Bates, H. W. (1862). "Contributions to an Insect Fauna of the Amazon Valley. Lepidoptera: *Helaconidae*." *Transactions of the Linnean Society* 23: 495–566. (TLS)

Brower, J. V. Z. (1958). "Experimental Studies of Mimicry in Some North American Butterflies. Part I. The Monarch, *Danaus flexippus*, and Viceroy, *Limenitis arcippus archippus*." *Evolution* 12:32–47.

Brower, L. P., J. V. Z. Brower, and C. T. Collins. (1963). "Experimental Studies of Mimicry. Relative Palatability and Müllerian Mimicry Among Neotropical Butterflies of the Subfamily *Heliconiinae*." *Zoologica* 48:65–83.

Darwin, C. D. (1863). "A Review of Mr. Bates' Paper on 'Mimetic Butterflies.'" *Natural History Review*: 219–224.

O'Hara, J. E. "(1995). Henry Walter Bates-His Life and Contributions to Biology." *Archives of Natural History* 22: 195–219.

Pfennig, D., W. R. Harcombe, and K. S. Pfenning. (2001). "Frequency-Dependent Batesian Mimicry." *Nature* 410: 323.

Wallace, A. R. (1866). "Natural Selection." *The Athenaeum*, December 1, no. 2040:716–717.

SOURCES OF QUOTES

1. p.52 "I rose generally with the sun ...," NORA, p. 269.
2. p.53 "...twelve months elapsed," *ibid.*, p. 271.
3. p.53 "We were about twenty persons in all...," *ibid.*, p. 203
4. p.54 "On the evening of the third ...," *ibid.*, pp. 388–389.
5. p.55 "I think I have got a glimpse into the laboratory," *Correspondence* vol. 9, p. 74.
6. p.59 "...the resemblance ...," TLS, p. 510.
7. p.59 "The explanation of this ...," *ibid.*, p. 511.

8.   p.60 "...one of the most remarkable...," *Correspondence*, vol. 10 (November 20, 1862).

9.   p.62 "My criticisms may be condensed ...," *Correspondence*, vol. 11 (April 18, 1863).

10.   p.62 "It may be said therefore ...," NORA, p. 353.

Chapter 4
Java Man

BOOKS

Darwin, Charles. (1871). *The Descent of Man and Selection in Relation to Sex.* London: John Murray.

Haeckel, Ernst. (1887). *The History of Creation: Or the Development of the Earth and Its Inhabitants by the Action of Natural Causes.* Translation revised by E. Ray Lankester, D., New York: Appleton Company.

Huxley, Thomas H. (1959). *Man's Place in Nature.* Ann Arbor: The University of Michigan Press (reprint of 1863 edition).

*Shipman, Pat. (2001). *The Man Who Found the Missing Link: Eugène Dubois and His Lifelong Quest to Prove Darwin Right.* New York: Simon and Schuster.

Theunissen, Bert. (1989). *Eugène Dubois and the Ape-Man from Java: The History of the First "Missing Link" and Its Discoverer.* Dardrecht, the Netherlands: Kluwer Academic Publishers.

ARTICLES

de Vos, John. (2004). "The Dubois Collection: A New Look at an Old Collection." In *VII International Symposium Cultural Heritage in Geosciences, Mining, and Metallurgy: Libraries–Archives–Museums: Museums and Their Collections. Scipla Geologic*, Special Issue 4:267–285.

SOURCES OF QUOTES

1.   p.68 "The question of questions for mankind...," *Man's Place in Nature*, p. 71.

2.   p.69 "...let us... disconnect our thinking selves...," *ibid.*, p. 85.

3.   p.70 "...is man so different...," *ibid.*, pp. 85–86.

4.   p.70 "Being happily free from all real...," *ibid.*, p. 86.

5.   p.70 "...the two pairs of limbs and...," *The History of Creation*, p. 299.

6. p.70 "Speechless man (Alalus)...," *ibid.*, p. 300.

7. p.71 "Where then must we look for primaevel Man?" *Man's Place in Nature*, p. 184.

8. p.75 "Everything here has gone against me...," *Eugène Dubois and the Ape-Man*, p. 40.

9. p.79 "I discover there is no more unsuitable place...," *The Man Who Found the Missing Link*, p. 159.

10. p.81 "*Pithecanthropus erectus* is the transitional form...," Dubois, Eugène. *Pithecanthropus erectus: Eine menschen-aehnliche Vebergangstorm aus Java. Batavia*, 1894. (cited in *The Man Who Found the Missing Link*, p. 209).

11. p.85 "...able discoverer of *Pithecanthropus*," Haeckel, Ernst. "On Our Present Knowledge of the Origin of Man." *Annual Report of the Board of Regents of the Smithsonian Institution for the Year Ending June 30, 1898*. Translation of a discourse given at the Fourth International Congress of Zoology at Cambridge, England (1898). (cited in *The Man Who Found the Missing Link*, pp. 306–307).

Chapter 5
Where the Dragon Laid Her Eggs

BOOKS

Andrews, Roy Chapman. (1926). *On the Trail of Ancient Man: A Narrative of the Field Work of the Central Asiatic Expeditions*. New York: Putnam.

Andrews, Roy Chapman. (1932). *The New Conquest of Central Asia: A Narrative of the Explorations of the Central Asiatic Expeditions in Mongolia and China, 1921–1930*. New York: American Museum of Natural History.

*Andrews, Roy Chapman. (1943). *Under a Lucky Star, A Lifetime of Adventure*. New York: Viking Press.

Bausum, Ann. (2000). *Dragon Bones and Dinosaur Eggs: A Photobiography of Explorer Roy Chapman Andrews*. Washington DC: National Geographic Society.

*Gallenkamp, Charles. (2001). *Dragon Hunter: Roy Chapman Andrews and the Central Asiatic Expeditions*. New York: Viking Press.

Rexer, Lyle. (1995). *American Museum of Natural History: 125 Years of Expedition and Discovery*. New York: H. N. Abrams (in association with the American Museum of Natural History).

SOURCES OF QUOTES

1. p.90 "...so you can get 'em all next time," *Under a Lucky Star*, p. 12.
2. p.91 "I'm not asking for a position...," *ibid.*, p. 22.
3. p.93 "...at last, there it was spreading its length like a slumbering gray serpent...," *ibid.*, p. 116.
4. p.94 "Never again will I have such a feeling as Mongolia gave me...," *Dragon Hunter*, p. 73.
5. p.94 "...should try to reconstruct the whole past history of the Central Asian plateau...," *Under a Lucky Star*, p. 163.
6. p.95 "Roy, we've got to do it. This plan...," *ibid.*, p. 164.
7. p.95 "It's a great plan; a great plan...," *ibid.*, p. 167.
8. p.97 "...geology was all obscured by sand," *The New Conquest of Central Asia*, p. 7.
9. p.98 "Well, Roy, we've done it. The stuff is here...," *On the Trail of Ancient Man*, p. 78.
10. p.99 "Slowly I became conscious that the air was vibrating...," *Under a Lucky Star*, p. 191.
11. p.101 "...paved with white fossil bones and all represented animals unknown," *On the Trail of Ancient Man*, p. 216.
12. p.102 "You have written a new chapter...," *Under a Lucky Star*, p. 200.
13. p.102 "Then our indifference evaporated. It was certain they really *were* eggs...," *Under a Lucky Star*, p. 213 and *On the Trail of Ancient Man*, pp. 228–229.
14. p.103 "Nothing in the world was further from our minds," *Under a Lucky Star*, p. 215.
15. p.103 "While the rest of us were on our hands and knees...," *ibid.*, p. 213.
16. p.104 "Let's keep the flour for work," *On the Trail of Ancient Man*, p. 232.
17. p.105 "...an unidentifiable reptile," *ibid.*, p. 327.
18. p.105 "Do your utmost to get some other skulls," *ibid.*, p. 328.
19. p.105 "Well I guess that's an order...," *ibid.*, p. 328.
20. p.106 "...possibly the most valuable seven days of work...," *ibid.*, p. 329.
21. p.107 "Dear God, my tent is full of snakes," Andrews, R. C. *This Business of Exploring*. New York: G. P. Putnam's Sons, 1935, p. 41.
22. p.107 "...fight its way across the long miles of desert...," *The New Conquest of Central Asia*, p. 310.

23. p.108 "...always intended to be an explorer...," *Under a Lucky Star*, p. 13.
24. p.108 "Always there has been an adventure...," *ibid.*, p. 300.

Chapter 6
The Day the Mesozoic Died

BOOKS

Alvarez, Luis W. (1987). *Alvarez: Adventures of a Physicist*. New York: Basic Books.
*Alvarez, Walter. (1997). *T. Rex and the Crater of Doom*. Princeton, NJ: Princeton University Press.
Powell, James L. (1998). *Night Comes to the Cretaceous: Dinosaur Extinction and the Transformation of Modern Geology*. New York: Freeman.

ARTICLES

Alvarez, L. (1983). "Experimental Evidence That an Asteroid Impact Led to the Extinction of Many Species 65 Million Years Ago. *Proceedings of the National Academy Sciences* 80:627–642.
Alvarez, L., et al. (1980). "Extraterrestrial Cause for the Cretaceous-Tertiary Extinction: Experimental Results and Theoretical Interpretation." *Science* 208:1095–1108.
Alvarez, W., et al. (1990). "Iridium Profile for 10 Million Years Across the Cretaceous-Tertiary Boundary at Gubbio (Italy)." *Science* 250:1700–1702.
Claeys, P., et al. (2002). "Distribution of Chicxulub Ejecta at the Cretaceous-Tertiary Boundary." *Geological Society of America*, Special Paper 356:55–68.
Hildebrand, A. R. (1991). "Chicxulub Crater: A Possible Cretaceous/Tertiary Boundary Impact Crater on the Yucatán Peninsula, Mexico." *Geology* 19:867–871.
Kring, D., and D. Durda. (2003). "The Day the World Burned." *Scientific American* 289:98–105.
Mukhopadhyay, S., et al. (2001). "A Short Duration of the Cretaceous-Tertiary Boundary Event: Evidence from Extraterrestrial Helium-3." *Science* 291:1952–1955.
Orth, C. J., et al. (1981). "An Iridium Abundance Anomaly at the Palynological Cretaceous-Tertiary Boundary in Northern New Mexico." *Science* 214:1341–1343.

Pope, K. O. (1991). "Mexican Site for K/T Impact Crater?" *Nature*: Scientific Correspondence 351:105.

Pope, K. O., et al. (1998). "Meteorite Impact and the Mass Extinction of Species at the Cretaceous/Tertiary Boundary." *Proceedings of the National Academy of Sciences* 95:11028–11029.

Schuraytz, B. C., et al. (1996). "Iridium Metal in Chicxulub Impact Melt: Forensic Chemistry on the K-T Smoking Gun." *Science* 271:1573–1576.

Smit, J. (1999). "The Global Stratigraphy of the Cretaceous-Tertiary Boundary Impact Ejecta." *Annual Review of Earth Planet Sciences* 27:75–113.

Simonson, B. M., and B. P. Glass. (2004). "Spherule Layers—Records of Ancient Impacts." *Annual Review of Earth Planet Sciences* 32:329–361.

WEBSITES

From Nobel Lectures, Physics, 1963–1970. "Luis Alvarez: The Nobel Prize in Physics 1968, Biography." Elsevier Publishing Company, Amsterdam, 1972. http://nobelprize.org/nobel_prizes/physics/laureates/1968/alvarez-bio.html

The Department of Paleobiology, Smithsonian National Museum of Natural History. "Tiny Creatures Tell a Big Story: Blast From the Past!–Part 2." http://paleobiology.si.edu/blastPast/paleoBlast2.html

SOURCE OF QUOTES

1. p.120 "...something unpleasant had happened to the Danish sea bottom," *T. Rex and the Crater of Doom*, p. 70.

Chapter 7
Miss Latimer's Extraordinary Fish

BOOKS

Smith, J. L. B. (1956). *Old Fourlegs: The Story of the Coelacanth*. New York: Longman, Green.

Thomson, K. S. (1991). *Living Fossil: The Story of the Coelacanth*. New York: W. W. Norton.

*Weinberg, S. (2000). *A Fish Caught in Time*. New York: Harper Collins.

ARTICLES

Colbert, E. H. (1939). "A Fossil Comes to Life." *Natural History* (May).

Courtenay-Latimer, M. (1989) "Reminiscences of the Discovery of the Coelacanth, *Latimeria chalumnae*." *Cryptozoology* 8:1–11.

Courtenay-Latimer, M. (1979). "My Story of the First Coelacanth." *Occidental Papers of the California Academy of Sciences* 134: 6–10.

Erdman, M. V., R. L. Caldwell, et al. (1998). "King of the Sea Discovered." *Nature* 395: 335.

Holder, M. T., M. V. Erdman, et al. (1999). "Two Living Species of Coelacanths?" *Proceedings of the National Academy of Sciences U.S.A.* 96: 12616–12620.

Smith, J. L. B. (1939) "A Living Fish of Mesozoic Type." *Nature* 143: 455–456.

Smith, J. L. B. (1953) "The Second Coelacanth." *Nature* 171: 99–101.

SOURCES OF QUOTES

1. p.133 "Oh dear, so much still to do," "Reminiscences," p. 4.
2. p.134 "...the most beautiful fish I had ever seen," "My Story," p. 7.
3. p.134 "Ah well, they're all asleep," *A Fish Caught in Time*, p. 4.
4. p.135 "...a bomb seemed to burst in my brain," *ibid.*, p. 15.
5. p.135 "MOST IMPORTANT PRESERVED SKELETON...," "My Story," p. 8.
6. p.135 "Although I had come prepared, that first sight...," *A Fish Caught in Time*, p. 20.
7. p.141 "Fourteen of the best years of my life...," *ibid.*, p. 64.
8. p.142 "It was the biggest mistake I ever made," *ibid.*, 171.
9. p.144 "This story is one of the most astounding...," "My Story," p. 10.

Chapter 8
A Sickle-Cell Safari

Much of the biographical background for this story was obtained by an interview with Tony Allison by telephone on June 1, 2007 and personal communications from Tony Allison to the author on June 7, 2007 and July 5, 2007.

ARTICLES

Aidoo M., D. J. Terlouw, et. al. (2002). "Protective Effects of the Sickle Cell Gene Against Malaria Morbidity and Mortality." *Lancet* 359: 1311–1312.

Allison, A. C. (1954). "Notes on Sickle-Cell Polymorphism." *Annals of Human Genetics* 19: 39–51.

Allison, A. C. (1954). "The Distribution of the Sickle-Cell Trait in East Africa and Elsewhere, and its Apparent Relationship to the Incidence of Subtertian Malaria." *Transactions of the Royal Society of Tropical Medicine and Hygiene* 48: 312–318.

Allison, A. C. (1954). "Protection Afforded by Sickle-Cell Trait Against Subtertian Malarial Infection." *British Medical Journal* 1: 290–294.

Allison, A. C. (1954). "Two Lessons from the Interface of Genetics and Medicine." *Genetics* 166: 1591–1599.

Allison, A. C. (1955). "Aspects of Polymorphism in Man." *Cold Spring Harbor Symposia on Quantitative Biology* 20: 291–303.

*Allison, A. C. (1956). "Sickle Cells and Evolution." *Scientific American* 195: 87–94.

Allison, A. C. (1957). "Properties of Sickle-Cell Haemoglobin." *Biochemical Journal* 65: 212–219.

Allison, A. C. (2002). "The Discovery of Resistance to Malaria of Sickle-Cell Heterozygotes." *Biochemistry and Molecular Biology Education* 30: 279–287.

Allison, A. C., and D. F. Clyde. (1961). "Malaria in African Children with Deficient Erythrocyte glucose-6-phosphate Dehydrogenase." *British Medical Journal* 1: 1346–1349.

Allison, A. C., and E. W. Ikin, et. al. (1952). "Blood Groups in Some East African Tribes." *Journal Royal Anthropological Institute of Great Britain and Ireland* 82: 55–61.

Allison, A. C., E. W. Ikin, and A. E. Mourant. (1954). "Further Observations on Blood Groups in East African Tribes." *Royal Anthropological Institute of Great Britain and Ireland* 84: 158–162.

Smith, S. M. (1954). "Appendix to Notes on Sickle-Cell Polymorphism." *Annals of Human Genetics* 19: 51–57.

Chapter 9
In Cold Blood: The Tale of the Icefish

BOOKS

Norske videnskaps-akademi i Oslo. (1935). *Scientific Results of the Norwegian Antarctic Expeditions*, 1927–1928. Olaf Holtedahl, ed. Oslo: I kommisjon hos J. Dybwad.

ARTICLES

Chen, Liangbiao, Arthur L. DeVries, and Chi-Hing C. Cheng. (1997). "Evolution of Antifreeze Glycoprotein Gene from a Trypsinogen Gene in Antarctic Notothenioid Fish." *Proceedings of the National Academy of Sciences U.S.A.* 94: 3811–3816.

Cheng, Chi-Hing C., and Chen, Liangbiao. (1999). "Evolution of an Antifreeze Glycoprotein." *Nature* 401: 443–444.

Cocca, Ennio, Manoja Ratnayake-Lecamwasam, et. al. (1995). "Genomic Remnants of α-globin in the Hemoglobinless Antarctic Icefishes." *Proceedings of the National Academy of Sciences U.S.A.* 92: 1817–1821.

DeVries, Arthur L (1971). "Glycoproteins as Biological Antifreeze Agents in Antarctic Fishes." *Science* 172: 1152–1155.

DeVries, Arthur L., and Donald E. Wohlschlag. (1969). "Freezing Resistance in Some Antarctic Fishes." *Science* 163:1073–1075.

di Prisco, Guido, Ennio Cocca, et. al. (2002). "Tracking the Evolutionary Loss of Hemoglobin Expression by the White-Blooded Antarctic Icefishes." *Gene* 295: 185–191.

Eastman, Joseph T., and Arthur L. DeVries. (1986). "Antarctic Fishes." *Scientific American* 254: 106–114.

Goodman, Billy. (1998). "Where Ice Isn't Nice." *BioScience* 48: 586–590.

Logsdon, John M., and W. Ford Doolittle. (1997). "Origin of Antifreeze Protein Genes: A Cool Tale in Molecular Evolution." *Proceedings of the National Academy of Sciences U.S.A.* 94: 3485–4687.

Near, Thomas J. (2004). "Estimating Divergence Times of Notothenioid Fishes Using a Fossil-Calibrated Molecular Clock." *Antarctic Science* 16.1: 37–44.

Ruud, Johan T. (1954). "Vertebrates without Erythrocytes and Blood Pigment." *Nature* 173: 848–850.

*Ruud, Johan T. (1965). "The Ice Fish." *Scientific American* 213: 108–115.

Sidell, Bruce D., and Kristin M. O'Brien. (2006). "When Bad Things Happen to Good Fish: The Loss of Hemoglobin and Myoglobin Expression in Antarctic Icefishes." *Journal of Experimental Biology* 209: 1791–1802.

Sidell, Bruce D., Michael E. Vayda, et. al. (1997). "Variable Expression of Myoglobin among the Hemoglobinless Antarctic Icefishes." *Proceedings of the National Academy of Sciences U.S.A.* 94: 3420–3424.

Somero, George N., and Arthur L. DeVries (1967). "Temperature Tolerance of Some Antarctic Fishes." *Science* 156: 257–258.

SOURCE OF QUOTES

1. p.170 "Do you know there are fishes here...Please bring some back...I have seen such a fish," *The Ice Fish*, p. 108.

Index